Rudolph / Kiesel / Aumnate

Understanding Plastics Recycling

A plus for you – additional digital content!

On our download portal you will find free additional material relating to this title.
Simply enter this code

plus-83hz4-t2b92

at the following web page:

plus.hanser-fachbuch.de/en

Natalie Rudolph
Raphael Kiesel
Chuanchom Aumnate

Understanding Plastics Recycling

Economic, Ecological, and Technical Aspects of Plastic Waste Handling

2nd Edition

HANSER

Hanser Publishers, Munich

Hanser Publications, Cincinnati

The Authors:

Natalie Rudolph, Ph.D., Netzsch Gerätebau GmbH, Selb, Germany

Raphael Kiesel, M.Sc. M.Sc., Fraunhofer Institute for Production Technology (IPT), Aachen, Germany

Chuanchom Aumnate, Ph.D., Metallurgy and Materials Science Research Institute, Chulalongkorn University, Bangkok, Thailand

Distributed in the Americas by:
Hanser Publications
414 Walnut Street, Cincinnati, OH 45202 USA
Phone: (800) 950-8977
www.hanserpublications.com

Distributed in all other countries by:
Carl Hanser Verlag
Postfach 86 04 20, 81631 Munich, Germany
Fax: +49 (89) 98 48 09
www.hanser-fachbuch.de

Library of Congress Control Number: 2020942248

Editor: Dr. Julia Diaz Luque
Production Management: Der Buchmacher, Arthur Lenner, Windach
Coverconcept: Marc Müller-Bremer, www.rebranding.de, Munich
Coverdesign: Max Kostopoulos
Typesetting: Kösel Media, Krugzell, Germany
Printed and bound by Druckerei Hubert & Co GmbH und Co KG BuchPartner, Göttingen
Printed in Germany

ISBN: 978-1-56990-846-4
E-Book ISBN: 978-1-56990-847-1

Foreword

Plastics are everywhere in our lives these days and accompany us throughout the day: the toothbrush and toothpaste in the morning; the windscreen wiper, seat, or window lifter of the car on the way to work; the keyboard at work; or the wrapping film over our vegetables at the grocery store. As these examples show, plastics have very different purposes and uses. While they often save us weight in technical applications and thus reduce fuel consumption, there are also many everyday objects whose use should be reconsidered again and again with regard to their entire life cycle. If, for example, the function, weight, or durability has been improved, their use is often advantageous due to the positive influence on the entire life cycle. However, if the use has more negative than positive consequences, alternatives should be considered already in the design phase. Despite the countless innovations that have been made possible by plastics and will continue to be realized in the future, the sustainable use of this valuable material is indispensable.

If the use of plastic proves to be the best option, reuse and repair should be considered as the next option. We are a long way from this in industrial use, but innovative solutions are also conceivable here. According to the motto "reduce, reuse, recycle", the recycling of plastic waste should only be the third option. This does not mean, however, that recycling is unimportant. The recycling of plastic waste is gaining in importance day by day and has now also come into the focus of the general public. This is mainly due to the alarming figures for plastic waste in the oceans. At present, more than 8 million tons of plastics are discharged into the oceans every year–and this number will increase if we do not change the way we handle plastic waste. Awareness of these catastrophic effects has already led to a change in public thinking and the use of plastic bags when shopping is now as absurd as disposing of a toothbrush after a single use.

For this reason, our book shows what unused potential lies in the recycling of plastics–from an ecological, economic, and technological point of view. Our focus is on the recycling of packaging waste. Plastics currently represent a great challenge–especially for the environment–and their recycling offers all the more opportunities. In addition, the non-reuse of plastics is equivalent to the loss of crude oil and is therefore also considered from this point of view.

To illustrate this potential, the book starts with a general overview of waste treatment strategies for plastics in the United States, and discusses the importance of plastic waste and some insights into how consumer behavior could be positively affected (Chapters 1 and 2). Chapter 3 focuses on the technical aspects and different processes of plastics recycling. In separate chapters, the economic (Chapter 4) and ecological properties (Chapter 5) of different waste treatment strategies for plastics are compared and evaluated. The analysis shows the potential of plastics recycling and the necessary boundary conditions for an increase in the recycling rate. Therefore, different scenarios for increasing the profitability of recycling are analyzed in Chapter 6. Last but not least, Chapter 7 presents the global potential for waste treatment and, in particular, plastics recycling using the examples of Europe and China.

We hope that with our book we can show you the importance and opportunities that the recycling of plastics offers and how we can all together play a role in making our world a little bit better—whether as decision-makers in a large company, when doing your weekly shopping in the supermarket, or when disposing of waste. Because, as you will discover in the course of reading this book: even the small things can have a huge effect!

We would like to thank everyone who supported us in writing and extending the second edition of this book. Special thanks to Sebastian Goris for adding his expertise in the area of fiber-reinforced plastics to the third chapter of this book.

The Authors

Aachen/Bangkok/Selb, July 2020

Contents

Acronyms and Other Abbreviations

Abbreviation	Description
ABS	acrylonitrile butadiene styrene
ARR	average rate of return
ASTM	American Society for Testing and Materials
CCM	cost comparison method
CLF	closed loop fund
DSC	differential scanning calorimetry
EPA	U.S. Environmental Protection Agency
EPS	expanded polystyrene
GHG	greenhouse gas
HDPE	high-density polyethylene
HIPS	high-impact polystyrene
LDPE	low-density polyethylene
LFG	landfill gas
LLDPE	linear low-density polyethylene
MFI	melt flow index
MFR	melt flow rate
MRF	materials recovery facility
MSW	municipal solid waste
OCC	old corrugated cardboard
PA	polyamide
PBT	polybutylene terephthalate
PC	polycarbonate
PCM	profit comparison method
PE	polyethylene
PEEK	polyether ether ketone (or polyarylether etherketone)
PET	polyethylene terephthalate
PLA	polylactide
PMMA	polymethyl methacrylate
POM	polyoxymethylene (polyacetals)

Abbreviation	Description
PP	polypropylene
PPE	polyphenylene ether
PPP	purchasing power parity
PRF	plastics recycling facility
PS	polystyrene
PTFE	polytetrafluoroethylene
PU	polyurethane
PVC	polyvinyl chloride
QA/QC	quality assurance/quality control
RCRA	Resource Conservation and Recovery Act
rLDPE	recycled low-density polyethylene
RoM	rule of mixtures
rPP	recycled polypropylene
SAN	styrene acrylonitrile
SNCR	selective noncatalytic reduction
SPP	static payback period
UV	ultraviolet
WARM	EPA Waste Reduction Model
WTE	waste-to-energy
XPS	extruded polystyrene

1 All About the Waste

Empty plastic bottles, yesterday's newspapers, eggshells, or used tea bags–all of these have one thing in common: they belong to *municipal solid waste (MSW)*, ordinarily called "trash" or "garbage". MSW refers to household, office, or retail waste and is part of everyone's daily life. Each person in the United States generates 2 kg of MSW every day, which resulted in a total waste creation of 268 million tons (t)[1] in 2017, and the trend is rising. As the volume of waste produced in the United States continues to increase, the handling and disposal of MSW are growing concerns of society. [1, 2, 3] In order to understand the impact of plastics and their recycling in the waste stream, it is important to understand the composition of MSW and current strategies for handling it.

1.1 Municipal Solid Waste–A Daily Companion

Municipal solid waste (MSW) is generally defined as nonhazardous waste. The United States Environmental Protection Agency (EPA) characterizes MSW as *"waste consisting of everyday items, used and then thrown away, such as product packaging, bottles, food scraps and newspapers which comes from homes, schools, hospitals and businesses."* MSW is classified in three broad categories according to where it is generated: household waste, commercial waste, and institutional waste. [4, 5] Household waste includes that from single and multiple family homes, hotels and motels, and day-use recreation areas. This category accounts for 55 to 65% of the total MSW generated and is also referred to as postconsumer waste. Thus, households are the primary source of MSW in the United States. Commercial waste includes solid waste from stores, offices, restaurants, warehouses, and other nonmanufacturing activities. Waste from schools, colleges, and similar public or

[1] Unless otherwise stated, "ton" in this book always refers to a metric ton (1000 kg).

quasi-public buildings is listed as institutional waste. The waste generated by the industrial sector is negligible. Manufacturing companies manage their solid residues or preconsumer waste by recycling, direct reuse, or self-disposal in industrial waste landfills. [1, 4, 5]

The total MSW generation in the United States increased steadily between 1960 and 2017, from 88 million t/year (tons per year) to 268 million t/year, due to economic expansion. The generation per capita per year has risen from 1.20 kg to 2.05 kg. However, it is important to note that after a peak in 2000, the MSW produced per capita per year has decreased slightly since then (2.15 kg to 2.05 kg). [2] Factors affecting the quantity of MSW generated include changes in population, individual purchasing power, product packaging, and technology, which again affect disposal habits as well as the nature of materials disposed. Hence, an analysis of the MSW composition over time is necessary to explain and also forecast future MSW generation in the United States. [4] The EPA uses two methods to characterize MSW, by material and by major products. In order to understand the studies presented in this book, knowledge of the proportion and quantity of different materials in MSW (Figure 1.1) is important and reflects the changes described in MSW generated in the United States.

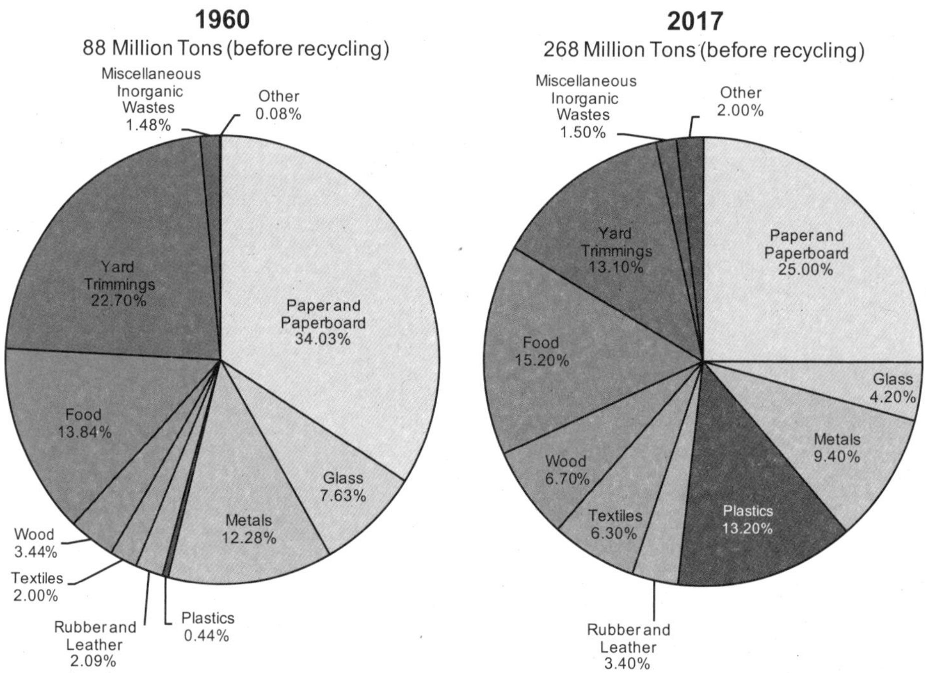

Figure 1.1 Comparison of the types and amounts of plastic waste generated in the United States in 1960 and 2017[2]

[2] The waste composition of 2017 is the basis for further calculations in this book.

Figure 1.1 provides information about the trends in consumer behavior and production of waste as seen, for example, in the significant change in the proportion of plastics in MSW between 1960 and 2017, which will be further discussed in Chapter 2. In this chapter we will discuss how the composition of MSW is of particular relevance for waste handling.

1.2 Management Methods for Municipal Solid Waste

The United States manages wastes in three different ways:

- Landfilling
- Incineration with energy recovery [waste-to-energy (WTE)]
- Recycling

Figure 1.2 shows a flowchart of this waste management system. Due to a growing awareness of environmental pollution, the EPA has strived to recycle as much MSW as possible or at least to burn it for energy, which will be further discussed in Chapter 3. It can be seen that residues from all waste treatment processes end up in landfill (dashed lines), which make disposal sites the final destination of MSW. [6, 7]

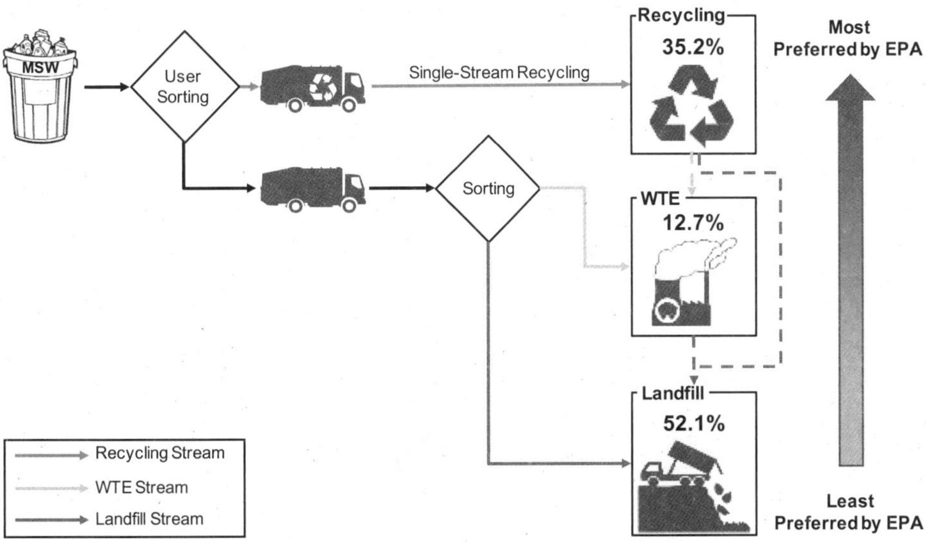

Figure 1.2 Flowchart of the waste management stream in the United States in 2017

Changing technologies, waste compositions (Section 1.1), and regulations, together with this growing environmental awareness, have induced a change in MSW treatment in recent decades. In 1960, 94% of the MSW generated in the United States was disposed of in landfills and afterwards open burned for volume reduction, which refers to burning garbage in outdoor pits. The remaining 6% was recycled and the amount of energy recovered was insignificant. Since then, the amount of waste that is burned for energy recovery or recycled has increased steadily. In 2017, the majority of waste was still disposed of in U.S. landfills, but compared to the 1960s, it was only 52.1%; 12.7% of MSW was burned with energy recovery and 35.2% was recycled (Figure 1.2). [1, 2, 4, 8]

1.2.1 Landfilling

Landfill describes engineered areas of land used for the controlled deposit of solid waste onto or into land. The EPA distinguishes between three sizes of landfills, which are listed in Table 1.1. [9]

Table 1.1 Classification of Landfill Sizes in the United States

	Weight [t]	Size [m^3]
Small Landfills	< 26,000	< 52,000
Medium Landfills	26,000–130,000	52,000–200,000
Large Landfills	> 130,000	> 200,000

United States MSW landfills are required to comply with federal regulations contained in subtitle D of the *Resource Conservation and Recovery Act (RCRA)*. The RCRA D requirements include siting restrictions in floodplains, surface and groundwater protection, disease and vector control, open-burning prohibitions, explosive gas (methane) control, fire prevention through the use of cover materials, and prevention of bird hazards to aircraft. [10, 11]

The United States mainly uses two different methods for active disposal of waste into landfills: the *area fill method* and the *trench method*. In the area fill method the waste is placed, spread out, and finally compacted in uniform layers using heavy equipment in large open sections of lined landfills. In the trench method, the waste is placed and compacted in a trench using material from the trench excavation as daily cover. The most appropriate method for a specific landfill is often determined by local conditions; a combination of both methods is possible. At the end of each day, cover materials are applied on top of the waste mass to prevent odors and fires, and to reduce litter, insects, and rodents. The daily cover material includes soil, compost, incinerator ash, foam, and tarps. If a landfill has reached its permitted height, the cell has to be closed. The required postclosure care period is 30 years, but it can be shortened or extended with changing state regulations and approvals. [4, 11, 12, 13, 14, 15]

Due to the described regulations, regularly amended by the EPA, opening a new or expanding an existing landfill plant has become increasingly difficult. After adoption of the first RCRA in 1968, the total number of landfills in the United States has steadily declined, especially small and private landfills. [1, 13]

Despite a decreasing number of landfills and a diminishing percentage of total MSW disposed of in landfills, the total waste capacity of landfills is still increasing due to the increasing amount of waste in general. The focus in the United States is on large and modern landfill plants. The fact that the combined capacity of the two largest landfill plants in the United States in 2017 was close to 10 billion t underlines this trend. [1, 13]

1.2.2 Incineration with Energy Recovery (Waste-to-Energy)

Incineration with energy recovery is the only *waste-to-energy (WTE)* technology that is commercially used in the United States. The heat generated by burning waste can be used directly for heating, to produce steam, or to produce electricity. In 2017, 12.7% of the total MSW in the United States was burned with energy recovery. [2, 16, 17]

The main motivation for burning waste was and still is the reduction of the volume and weight of solid waste. Nowadays, combustion decreases the volume of waste by about 90% and its weight by about 70%. [17]

The first garbage incinerator in the United States was built in 1885 by the U.S. Army. It only served to reduce the volume and weight of MSW and did not generate energy. At the beginning of the twentieth century, incineration became a convenient method to dispose of waste. Due to air emission problems, technical limitations, and the increasing popularity of landfills, the amount of waste being burned was negligible in the 1960s. Europe, however, constantly improved and built waste-to-energy technologies mainly because these countries had less land available for landfilling. Increasing oil prices in the 1970s, caused by the OAPEC oil embargo, as well as federal laws and policies finally forced the United States to build their own WTE incineration plants. [2, 17]

The facilities tended to be built near landfills of urban centers. On average, 1 ton of solid waste produces 525 kWh of electricity, which is equivalent to the energy produced by a $^{1}/_{4}$ ton of coal or 1 barrel of oil. [1, 17]

The most common type of incineration plant is the *mass-burn facility* (64 out of 84 of the facilities in the United States [18]). Therefore, this type of facility will be further considered in the analysis. Mass-burn facilities use solid waste directly out of the garbage truck, without shredding or processing of the material. A plant diagram of a mass-burn facility is shown in Figure 1.3.

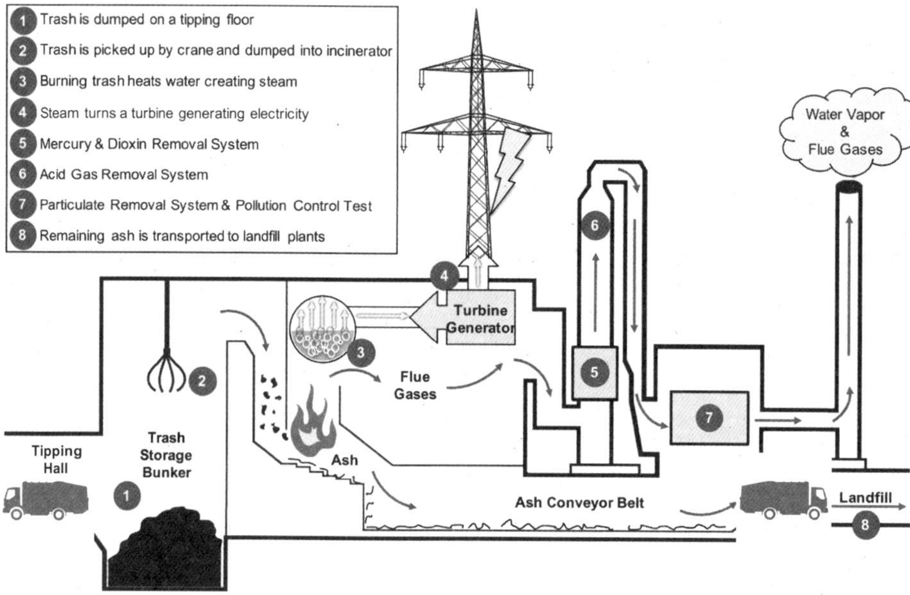

Figure 1.3 Waste-to-energy plant diagram of a mass-burn facility

After the trash is dumped on the tipping floor (1), it is picked up by a crane and dropped into an incinerator (2). Burning the trash in the incinerator generates three different "products": ash, flue gases, and heat. While both ash and flue gases are waste products, the generated heat is really valuable. The heat is used to boil water in a big tank to thereby create steam (3). The steam is used to run a turbine generator, which produces energy, mostly in form of electricity (4). The flue gases pass a pollution control system before they are mixed with water vapor and blown into the air. This control system includes a nitrogen oxide removal system, a mercury and dioxin removal system (5), an acid gas removal system (6), a particulate removal system (7), and the final pollution control test (7). The remaining ash is conveyed to a waste truck and transported to the closest landfill plant (8), where it is mixed with the untreated landfill waste. [17]

In 2014, 84 waste-to-energy facilities were in operation in the United States, with a combined capacity to generate 2,769 MWh of energy daily. [1, 18]

A fairly new process among the WTE technologies that converts plastics such as PE, PP, and PS into light crude oil is *pyrolysis*, or *thermochemical decomposition*. Also known as plastic-to-oil processing, it is able to produce 1 L of oil out of 1 kg of plastic, for which 1 kWh of electricity is required to operate the processing machinery. In the process, plastic flakes are heated to around 400 °C to decompose the plastic material into inactive carbon char and gas. The gas is then condensed and liquefied into oil. The remaining gas is filtered and released as carbon dioxide and steam. Although the technology is promising and commercially available, it is not yet economically viable due to the initial investment in most settings today. However, some successful prototype facilities exist in the United States and Canada. [19]

1.2.3 Recycling

Recycling converts materials and products that would otherwise become waste into valuable resources. Paper and cardboard, yard trimmings, metals, glass, and plastics are the most important recyclables in the United States. The main characteristic of the U.S. recycling process is single-stream recycling. Single-stream recycling means that all recyclables are picked up unsorted and are not separated until they get to the processing facility. [2]

The general cyclic recycling process involves three main steps:

1. Collecting the recyclables
2. Processing the recyclables and turning them into new products
3. Purchasing recycled products

After collecting the recyclables via curbside collection, drop-off programs, buyback operations, and container-deposit systems, they are transported to material recovery facilities (MRFs), mixed-waste processing facilities, or mixed-waste composting facilities. Since MRFs are the most common of these facilities in the United States and are more and more high tech, the analysis in this book will focus on this type of facilities (more details can be found in Chapter 4). [2, 5]

High-tech MRFs are characterized by the automated separation of unsorted recyclables using eddy currents, magnetic pulleys, optical sensors, and air classifiers, reducing manual sorting. Automatic sorting supports and simplifies single-stream recycling and enhances its economic profitability (more details can be found in Chapter 4). [2]

In 2017, 35.2% of the total MSW was recycled. Items with the highest recycling rates include lead-acid batteries (99.1%), steel cans (70.9%), paper and cardboard (65.9%), and aluminum cans (49.2%). Only 8.4% of the total plastic waste, which accounts for 13.2% of the total MSW, was recovered in 2017. Thus, the potential and motivation for improving the recycling of this widely used material is very high. [1, 20]

References

[1] Center for Sustainable Systems. *Municipal Solid Waste.* University of Michigan, Ann Arbor, MI. 2015.

[2] United States Environmental Protection Agency (EPA). *Advancing Sustainable Materials Management: 2016 and 2017 Tables and Figures.* Washington, D.C. 2019.

[3] United States Environmental Protection Agency (EPA). *Municipal Solid Waste in the United States: 2017 Fact Sheet.* Washington, D.C. 2019.

[4] United States Environmental Protection Agency (EPA). *Municipal Solid Waste Landfills–Economic Impact Analysis for the Proposed New Subpart to the New Source Performance Standards.* Washington, D.C. 2014.

[5] Shin, D. *Generation and Disposition of Municipal Solid Waste (MSW) in the United States–National Survey,* Master's Thesis, Columbia University. New York, NY. 2014.

[6] United States Environmental Protection Agency (EPA). *Advancing Sustainable Materials Management: 2013 Fact Sheet. Assessing Trends in Material Generation, Recycling and Disposal in the United States.* Washington D.C. 2015.

[7] The World Bank. *Decision Makers' Guide to Municipal Solid Waste Incineration.* The World Bank, Washington D.C. 1999.

[8] Klar, M., Gunnarsson, D., Prevodnik, A., Hedfors, C., and Dahl, U. *Everything you (don't) want to know about plastic.* Swedish Society for Nature Conservancy (Natrskyddsföreningen). 2014.

[9] Environmental Protection Authority (EPA), Australia. *Waste Guidelines.* Adelaide, Australia. 2009.

[10] United States Environmental Protection Agency (EPA). *Criteria for classification of solid waste disposal facilities and practices.* Washington, D.C. 2011.

[11] United States Senate. *Solid Waste Disposal Act (RCRA).* Washington, D.C. 2002.

[12] National Solid Wastes Management Association. *Modern Landfills: A Far Cry from the Past.* Washington, D.C. 2008.

[13] United States Environmental Protection Agency (EPA). *Regulatory Impact Analysis for the Proposed Revisions to the Emission Guidelines for Existing Sources and Supplemental Proposed New Source Performance Standards in the Municipal Solid Waste Landfills Sector.* Washington, D.C. 2015.

[14] Guyer, J.P. *Introduction to Sanitary Landfills.* CED Engineering, Stony Point, NY. 2009.

[15] Repa, E.W. *NSWMA's 2005 Tip Fee Survey.* National Solid Wastes Management Association, Washington D.C. 2005.

[16] Davis, J., Haase, S., and Warren, A. *Waste-to-Energy Evaluation: U.S. Virgin Islands.* National Renewable Energy Laboratory, Golden, CO. 2011.

[17] Texas Comptroller of Public Accounts. Municipal Solid Waste Combustion. In: *The Energy Report 2008.* Austin, TX. 2008. pp. 257–267.

[18] Michaels, T. *The 2014 ERC Directory of Waste-to-Energy Facilities.* Energy Recovery Council, Arlington, VA. 2014.

[19] Ignatyev, I.A., Thielemans, W., and Vander Beke, B. Recycling of polymers: a review. *ChemSusChem.* vol. 7, pp. 1579–1593, 2014.

[20] National Association of Counties. *Waste Energy Recovery: Renewable Energy from County Landfills.* Washington, D.C. 2015.

2 Plastics—Increasing Value, Decreasing Lifetime

"For the last 150 years, plastic materials have been key enablers for innovation and have contributed to the development and progress of society." [1] This quote from PlasticsEurope portrays the importance of plastics in today's industrialized society. Imagining today's world without plastics is virtually impossible. Compared to every other engineered material in the world, plastics have the highest growth rate. This is related to their outstanding properties: they are lightweight and easy and economical to manufacture into complex products. Plastics have become the most important raw material for a variety of products and applications from automotive components, mobile phones, pipes, and foams to packaging. Their versatility, durability, and high strength-to-weight ratio have led to a twentyfold exponential growth in their production in the past half-century. They constitute about 15% of the weight of a car and 50% of the Boeing Dreamliner. By 2030, their production is expected to double and by 2050, to quadruple. [2, 3, 4, 5, 6, 7]

Plastics' largest application is in packaging, representing 34% of total sales of plastics in the United States in 2012. Their low cost, light weight, and high performance make them suitable for packaging materials. In addition, plastic packaging benefits the environment in some ways: its low weight reduces fuel consumption for transportation and its barrier properties keep various foods fresh longer, thus leading to a reduction in food waste. As a result, plastics are progressively replacing other packaging materials. [5, 7]

This growth in plastics production and replacement of other materials in many industry sectors is leading to an increasing amount of plastic waste, as can be seen in Figure 2.1. Between 1960 and 2017, the total amount of plastic waste in the United States increased by a factor of 80. In parallel, the percentage of plastic waste in municipal solid waste (MSW) grew as well, which demonstrates the rising significance of plastics in our daily life. [3, 8] However, it should be noted that the rate of increase in the amount of plastic waste has slowed down slightly since 2005 in comparison to the steep increase between 1980 and 2005.

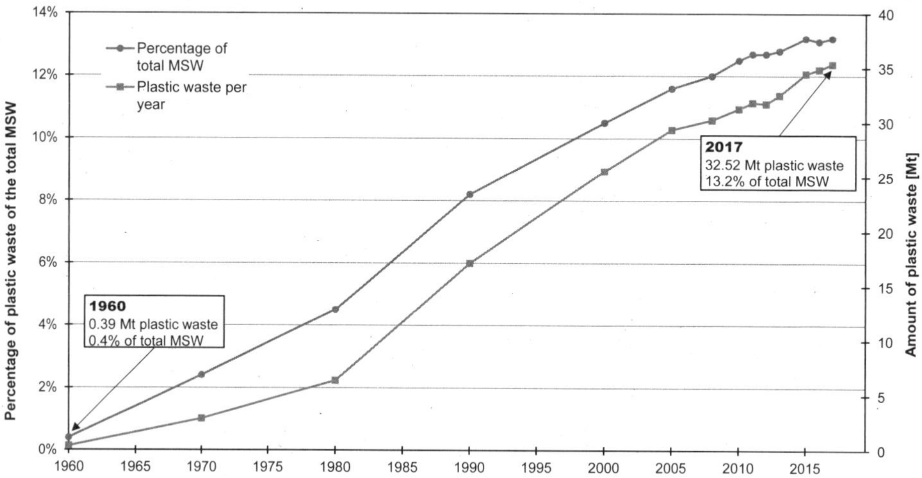

Figure 2.1 Changes in the amount of plastic waste in the U. S. between 1960 and 2017. MSW, municipal solid waste; Mt, megatons.

Comparing the amount of plastic waste to the total amount of plastic produced in the United States between 2005 and 2012, the ratio of waste to production increased from 26.8% in 2005 to 30% in 2012. The reason for this change is the reduced lifetime of plastic products caused by the increasing use of plastics for *single-use products*, such as plastic bags, packaging, and containers, which are immediately thrown away after unpacking the product they hold or contain. [3, 7, 8]

Polypropylene (PP), low-density polyethylene (LDPE), and *high-density polyethylene (HDPE)* are the three most common polymers found in the packaging waste stream, especially in the form of plastic films. Table 2.1 shows the breakdown of plastic waste by polymer. PP, LDPE, and HDPE add up to a total weight percentage of more than 60%. However, the recycling levels of plastic films from postconsumer waste are very small. [3, 9, 10]

Table 2.1 Plastic Waste by Polymer in the United States in 2017

Polymer	Generated		Recycled	
	Amount [kt]	Percentage of Plastic Waste [%]	Amount [kt]	Percentage of Plastic Waste [%]
PET	5,010	14.16	910	18.16
HDPE	6,150	17.39	580	9.43
PP	8,000	22.62	50	0.63
LDPE/LLDPE	8,080	22.84	340	4.21
PS	2,350	6.64	10	0.43
PVC	960	2.71	Negligible	–
Other resins	4,820	13.63	1,070	22.62
Total	**35,370**	**100.00**	**2,960**	**8.37**

Regardless of the positive aspects of the increasing use of plastics, the increasing plastic waste from single-use products generates significant negative consequences. The production of plastics consumes a large amount of the scarce resource, oil: the processing and raw materials for producing 1 kg of plastics require about 2 kg of crude oil. If the current plastic waste handling does not change, the ratio of plastic waste to fish (by weight) in the ocean in 2050 will be 1:1. Additionally, plastics' share of global oil consumption would increase from 6% (2012) to 20% (2050) and plastics' share of the carbon budget would rise from 1% (2012) to 15% (2050). Therefore, the emerging challenge of these developments from both an economic and an environmental point of view is to improve plastic waste handling and, in the best case, to recycle and reuse these resources. [5, 11]

Even though the topics of recycling, single-use packaging, and plastic waste in the environment have moved into the public eye, not all efforts to reduce and recycle waste on the individual level have the same success rate.

Different studies show that fines lead to more environmentally friendly behavior than incentives. For example, a fee of 5 cents for the use of disposable bags resulted in 45% of shoppers bringing their own reusable grocery bag, while the same 5 cents bonus generated no effect on the consumers' behavior. [12] Various studies by coffee shops on reusable cup usage have shown the same. [13, 14, 15, 16] Consumers are more motivated to avoid a fine, which is known as the model of loss aversion in behavioral economics. Early studies by Kahneman and Tversky [17] suggest that individuals perceive losses more strongly than gains. In their experiments, they posed two different problems to the individuals. One was phrased as a gain and one as a loss. In problem A they were given $1,000 and two options: a 50% chance to win another $1,000 or a gift of $500. Most individuals picked the certain gain, the gift of $500. In problem B they were given $2,000 and two options that were phrased as losses: a 50% chance to lose $1,000 or a certain loss of $500. Almost everyone now picked the gamble to potentially lose $1,000. The results of both problems were the same. In both cases the gamble gave people the 50% chance of getting $2,000. The certain option would give them $1,500. However, they perceived the pain to lose money more strongly than the pleasure to get more.

Four recent studies [16] further suggest that the framing as a fine versus an incentive signals people a social norm for this behavior. Two different scenarios–one with reusable shopping bags and one with reusable coffee mugs–were studied separately. In both cases, one group was told that a charge was imposed for not bringing a reusable cup or bag, while the other group was told that a discount was offered for bringing it. The first group inferred that bringing your item was more common and even more socially expected. The first groups commented further that they would feel more guilty for not bringing their own cup or bag.

In another of these four studies, the long term effect of surcharges was researched. It was found that fines have a more lasting effect than incentives and can therefore lead to lasting behavioral changes.

Instead of monetary fines or incentives, another research team [18] studied the effect of positive feedback on behavioral change. In a hospital the medical staff was asked to increase the frequency of hand sanitization to prevent the spreading of diseases. Even though the importance was stressed frequently and the staff was knowingly recorded, only 10% of the medical staff sanitized their hands before and after entering patients' rooms. Once an electronic board was introduced, which displayed a positive message such as "Good job!" as response to an employee sanitizing her/his hands, the compliance rose to close to 90% within one month and was more likely to be repeated in the future. The researchers concluded that this response is hardwired in our brains. They argue that to reap rewards in life, like a promotion, we have to take actions, and to avoid bad outcomes such as poison it is better to do nothing. This assumption is further supported by another function of our brain. We are more likely to believe and accept positive information than negative one. Both assumptions would explain why the positive, immediate feedback was more effective motivation than the threat of a probable illness.

Relating this to plastics recycling, it indicates that consumers will respond better to the promise of a cleaner environment when we reach our goals than to the threat of a polluted one if we do not.

Both the response to fines or a loss as well as to positive feedback are important aspects of our human behavior that should be taken into account when creating effective and actionable policies for plastics recycling and waste reduction.

References

[1] PlasticsEurope. *Plastics – the Facts 2014/2015. An analysis of European plastics production, demand and waste data.* Brussels, Belgium. 2015.

[2] Baur, E., Osswald, T. A., and Rudolph, N. *Plastics Handbook,* 5th ed. Hanser, Munich. 2019.

[3] United States Environmental Protection Agency (EPA). *Advancing Sustainable Materials Management: 2016 and 2017 Tables and Figures.* Washington, D. C. 2019.

[4] Kearney, A. T. *Plastics. The Future for Automakers and Chemical Companies.* 2012.

[5] Project MainStream. *The New Plastics Economy: Rethinking the Future of Plastics.* Ellen MacArthur Foundation, The World Economic Forum, and McKinsey & Company. 2016.

[6] Andrady, A. L., and Neal, M. A. Applications and societal benefits of plastics. *Philosophical Transactions of the Royal Society B: Biological Sciences.* vol. 364, pp. 1977–1984, 2009.

[7] American Chemistry Council, Economics & Statistics Department. *Plastic Resins in the United States.* Washington, D. C. 2013.

[8] United States Environmental Protection Agency (EPA). *Municipal Solid Waste in the United States: 2017 Fact Sheet.* Washington, D. C. 2019.

[9] Franklin Associates. *Impact of Plastics Packaging on Life Cycle Energy Consumption & Greenhouse Gas Emissions in the United States and Canada.* Prepared for American Chemistry Council (ACC) and Canadian Plastics Industry Association (CPIA). 2014.

[10] Moore Recycling Associates, Inc. *2014 National Postconsumer Plastic Bag & Film Recycling Report.* Sonoma, CA. 2014.

[11] Hopewell, J., Dvorak, R., and Kosior, E. Plastics recycling: challenges and opportunities. *Philosophical Transactions of the Royal Society of London B: Biological Sciences.* vol. 364, pp. 2115–2126, 2009.

[12] Homonoff, T. A. Can Small Incentives Have Large Effects? The Impact of Taxes versus Bonuses on Disposable Bag Use. *American Economic Journal: Economic Policy* [MOU1], 10 (4), pp. 177–210, 2018.

[13] N. N. *Starbucks Global Responsibility Report 2014.* Access Date: 2020/03/19. Available: *http://globalassets.starbucks.com/assets/ea2441eb7cf647bb8ce8bb40f75e267e.pdf*

[14] Sidhu, M., Mehrotra, K., and Hu, K. *Single-Use Items Reduction: Disposable Cups.* UBS Student research report, University of British Columbia. Access Date: 2020/03/19. Available: *https://sustain.ubc.ca/sites/default/files/seedslibrary/ECON492_SingleUseItemsReductionCoffeeCups_FinalReport.pdf*

[15] N. N. *Starbucks Rolls Out 5-cent Paper Cup Charge Across Germany, 2020.* Access Date: 2020/03/19. Available: *https://stories.starbucks.com/emea/stories/2020/starbucks-rolls-out-5-cent-paper-cup-charge-across-germany/*

[16] Lieberman, A., and Duke, K. *Why we're incentivized by discounts and surcharges.* Harvard Business Review. Access Date: 2020/03/19. Available: *https://hbr.org/2020/02/research-why-were-incentivized-by-discounts-and-surcharges*

[17] Tversky, A., and Kahneman, D. *Loss Aversion in Riskless Choice: A Reference-Dependent Model.* Quarterly Journal of Economics 106 (4), pp. 1039–61, 1991.

[18] Sharot, T. *What motivates employees more: rewards or punishments?* Harvard Business Review. Access Date: 2020/03/19. Available: *https://hbr.org/2017/09/what-motivates-employees-more-rewards-or-punishments*

3 Plastics Recycling—Conservation of Valuable Resources

Plastics recycling is the term used for reprocessing postconsumer and preconsumer plastic waste (manufacturing scrap) into useable products. The idea behind recycling is to break down finished products into their component materials and then use those materials as feedstock to manufacture new products. Based on the plastic waste source, the recycling process and the finished product differ. In general, plastics can only be reused a limited number of times before they are too degraded for further use. Currently, all preconsumer plastic waste is fed back into the plastic production stream, but only a little portion of postconsumer plastic waste is reclaimed for its original use. However, every bit of plastic that is recycled reduces the need for new plastic feedstock and thus decreases the amount of resources and energy used for its production.

Recycling of plastics for use in creating new high-quality plastic products requires that the recycled materials are clean and consist of only a single type of plastic. In such cases, the recycled plastic substitutes for virgin plastic. The big challenge in recycling postconsumer plastics, especially those from mixed (co-mingled and/or single) stream collection is that they are often contaminated. Recycling of mixed plastics is much more complicated. If the recycled plastics are contaminated and/or are a mixture of different types of plastic, the quality of the recycled plastic is lower; for example, the plastic may have lower strength. The challenge in managing the recycling of large quantities of a mixture of miscellaneous types of contaminated plastics needs to be considered using an integrated approach to source reduction, reuse, and recycling. [1]

Plastics recycling is more complex than metal or glass recycling because of the many different types of plastic. Thus, recyclability and environmental compatibility need to be criteria considered at the beginning of the design process of plastic products instead of as an afterthought, particularly in many products where several kinds of plastic and sometimes non-plastic components are integrated. The separation, recovery, and purification of the plastic components in such a product require several steps, which consume additional energy. Unfortunately, the recycling rate, the amount of any type of plastic that is recycled in a period of time, is directly related to the price of virgin resins for that type of plastic, which is related to the price of oil (see Section 4.4.5). Low oil prices result in low costs for the virgin resins. In these times, recycled resins are too expensive to be used by comparison, and the recycling rates drop. Therefore, the goal of any sustainable growth in recycling should be the maximization of efficiency of energy utilization in every step of the process, from the initial production of plastic goods to the disposal or recovery of plastic wastes. [2]

3.1 Plastics Recycling Methods

There are three common methods for plastics recycling: *mechanical recycling* (primary and secondary recycling) and *chemical recycling* (tertiary recycling). Based on the degree of contamination of the plastics (Section 3.5) with organic or inorganic substances (other polymers or impurities), one of these three recycling methods is chosen. The molecular structure of the plastics as well as existing cross-links, such as in thermosets or rubbers, also influence the decision process. [3, 4]

3.1.1 Mechanical Recycling

Among the recycling methods, mechanical recycling is the most desirable approach because of its low cost and high reliability. In general, mechanical recycling maintains the molecular structure of the polymer molecule. After grinding of the plastic waste material, the main processing step is remelting of the regrind material, which limits the use of mechanical recycling to thermoplastic polymers. Since remelting causes a degradation of the polymer chain, virgin material is often mixed with recycled material to reduce the effects of degradation on the product properties. The mixing leads to a dilution of the virgin material, which is described in Section 3.2.1.2. [5]

Mechanical recycling is divided into primary and secondary mechanical recycling, depending on whether the source of the waste is preconsumer or postconsumer, respectively. Preconsumer manufacturing scrap plastic is usually clean and of a single type or at least of a known composition and requires no further treatment, whereas postconsumer waste is highly contaminated and requires additional steps like collecting, sorting, and cleaning.

3.1.2 Chemical Recycling

Chemical recycling is used for *cross-linked polymers* or for thermoplastic polymers if no sufficient quality can be achieved using mechanical recycling. Chemical processes are used to convert the polymer chains to *low molecular weight* compounds or, in some cases, the original plastic monomer (feedstock). The monomers can be used for polymerization to generate the original polymer again, whereas the low molecular weight compounds are used as feedstock for the petrochemical industry. Common processes for this recycling method are hydrolysis, hydrocracking, pyrolysis, and depolymerization. Because of the large amounts of energy and chemicals consumed by the currently available processes, chemical recycling is only economically and ecologically reasonable for a very limited number of polymers such as polymethyl methacrylate (PMMA) and polyether ether ketone (PEEK). Chemical recycling of polyethylene terephthalate (PET) has been successfully developed. However, it is hindered by the processing cost. Furthermore, the chemical processing has been proven to be technically possible for polyolefins but is still in the laboratory stage of development. This is a fast growing research area, where significant breakthroughs can be expected in the next decade. [3, 4, 6, 7, 8]

■ 3.2 Recycling Different Types of Plastic Waste

As mentioned before, plastic waste can be divided into *preconsumer waste* (manufacturing scrap) and *postconsumer waste* (recovered waste). These different plastic waste types are recycled differently.

3.2.1 Preconsumer Waste

3.2.1.1 Manufacturing Scrap

Preconsumer waste, such as runners, gates, sprues, and trimming, is normally recycled using primary mechanical recycling. It is ground and remelted in-house.

3.2.1.2 Dilution Effect

Manufacturing scrap is often mixed into virgin material to reduce material cost while at the same time minimizing the effects of degradation on part performance. Depending on the mixing ratio, either the virgin material is diluted with regrind or the regrind is refreshed with virgin material. By using a constant mixing ratio during continuous processing, the regrind waste itself is diluted by material that has been reprocessed once, twice, three times, etc. The composition of a material with a proportion of recyclate q after n processing cycles can be calculated using Equation 3.1.

$$\sum_{i=1}^{n} q^{n-i}(1-q)=1 \qquad 3.1$$

For small proportions of recyclate, the regrind material contains only minimal amounts of material that has passed through a large number of processing cycles and therefore is highly degraded.

Figure 3.1 shows the composition of material with different mixing ratios of recycled and virgin material. The first column shows 30% recycled and 70% virgin material. Under these conditions, the regrind material contains less than 0.8% of material that has been reprocessed five times or more. Seventy percent of the material is virgin material, 21% has been processed once, 6.3% twice, and 1.9% three times. As proportions of material smaller than 1% do not have a significant influence on the material properties and can be neglected [9], the properties will be dominated by fractions that have been processed four times or less. Thus, it can be concluded that the properties of a material with small amounts of recyclate will not fall below a certain level. [10]

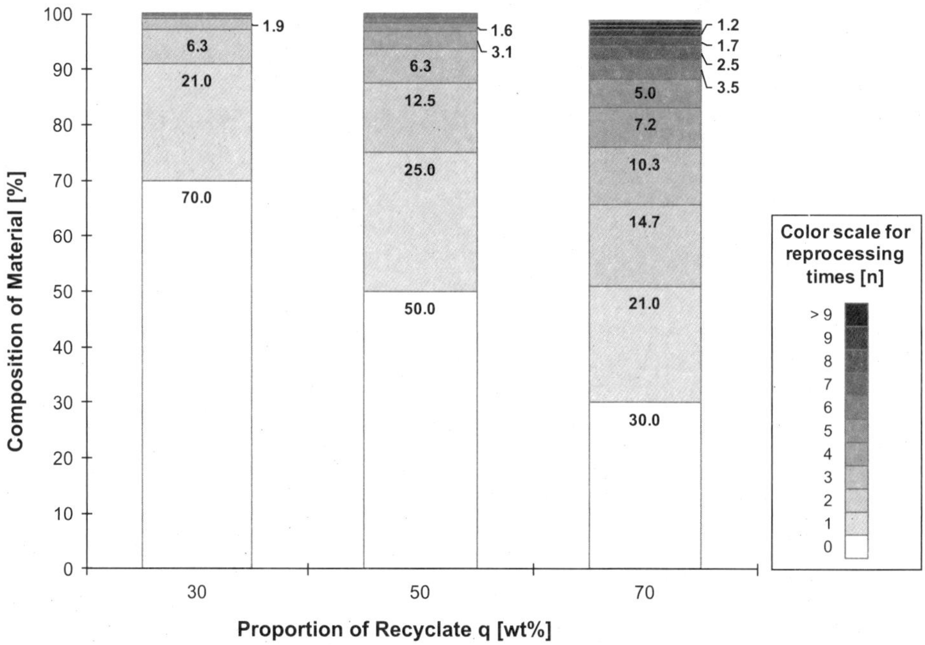

Figure 3.1 Composition of recycled plastic material after **n** reprocessing steps for 30%, 50%, and 70% recycled material

However, regrind material with high proportions of recyclate contains significant amounts of highly degraded material, as can be seen in the right column in Figure 3.1, in which 70% of the regrind is recycled and 30% is virgin material. This regrind material contains 5.0% material that has been reprocessed five times, as well as 30% that is virgin material, 21% that has been processed once, 14.7% twice, 10.3% three times, and 7.2% four times. After nine processing cycles, the material still contains 1.2% of the initial material. Although this mix contains significant portions of highly degraded material, after 10 reprocessing cycles the material reaches a steady state in which performance properties are not affected anymore by further processing. Therefore, this mixing ratio is used quite frequently for packaging products, e.g., PET containers.

3.2.2 Postconsumer Waste

Consumer plastics are largely made from six different polymer resins, which are indicated by a number, or *resin code*, from 1 to 7 molded or embossed onto the surface of the plastic product. The number 7 indicates any polymer other than those numbered 1 to 6. Table 3.1 lists the polymer resins, their resin codes, and the general applications for virgin and recycled plastics made from these resins. The percentages of the different types of postconsumer plastic waste in municipal solid waste (MSW) in the United States in 2017 are given in Table 2.1. [11]

Table 3.1 Plastic Types and Products from Virgin and Recycled Materials

Resin Symbol and Plastic Type	Products Created from Virgin Plastics	Products Created from Recycled Plastics
01 PET Polyethylene terephthalate	Bottles for water, soft drinks, salad dressing, peanut butter, and vegetable oil	Egg cartons, carpet, and fibers and fabric for T-shirts, fleeces, tote bags, shoes, etc.
02 HDPE High-density polyethylene	Milk and juice cartons, detergent containers, shower gel bottles, and shipping containers	Toys, pails, drums, traffic barrier cones, fencing, and trash cans
03 PVC Polyvinyl chloride	Packaging materials, plastic pipes, decking, wire and cable products, blood bags, and medical tubing	Shoe soles, construction material, and boating and docking bumpers

Table 3.1 Plastic Types and Products from Virgin and Recycled Materials *(continued)*

Resin Symbol and Plastic Type	Products Created from Virgin Plastics	Products Created from Recycled Plastics
04 LDPE Low-density polyethylene	Disposable diaper liners, cable sheathing, shrink-wrap, and film	Timbers, trash can liners, shopping envelopes, lumber, and floor tiles
05 PP Polypropylene	Medicine bottles, drinking straws, yogurt containers, butter and margarine tubs, automotive parts, and carpeting	Signal lights, bicycle racks, trays, battery cables, and ice scrapers
06 PS Polystyrene	Egg cartons, cups, food containers, plastic forks, and foam packaging	Egg cartons, foam packing, and light-switch plates
07 O All other resins or mixtures of resins	Mixed plastics or multilayer plastics packaging	–

The chemical composition and function of each resin controls where the resin can be recycled as well as the recycling rate. The latter is attributed to the difficulty of separating mixed plastic during the recycling process. For example, PET, or resin code 1, only accounts for 14.39% of the total plastic waste but it has the highest recycling rate of all resins. Because of its widespread use in transparent drinking bottles, PET is easy to identify and sort by transmission detectors.

Some resins are not compatible with others, because their molecular structures repel each other if mixed. This leads to deterioration of the mechanical performance of plastic products made from them if they are not engineered properly. Most plastics have additives incorporated to achieve certain additional properties such as flame retardancy, flexibility, or resistance to ultraviolet (UV) damage. This makes it nearly impossible to obtain a homogeneous plastic mixture with uniform behavior. Therefore, it is important that the sorting process is well regulated to ensure the integrity and overall performance of recycled plastic products.

Depending on their properties, different plastics are used for different applications. Currently, packaging, consumer and institutional products, and building and construction materials are the top three uses for plastics. The share of U.S. plastics demand by use in 2015 is shown in Figure 3.2. These different applications are again represented in the plastic waste. [12]

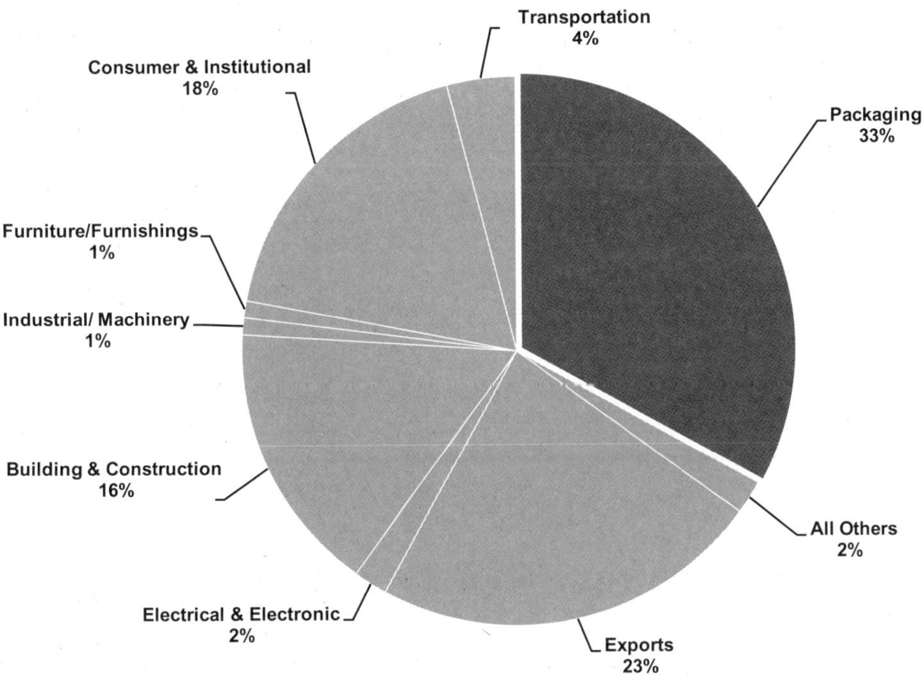

Figure 3.2 United States plastics demand by use in 2018 [11]

3.2.2.1 Packaging Plastic Waste

As a result of the demand for plastics, a large share of MSW plastics consists of packaging items in which high-density polyethylene (HDPE, 17.16%), low-density polyethylene (LDPE, 22.94%), and polypropylene (PP, 22.76%) together account for about 63% of the waste (see Table 2.1). The continuing increase in the use of disposable packaging has led to increasing amounts of plastics ending up in the waste stream. In Europe, where the proportion of plastic waste in the waste stream is similar to that in the United States, packaging waste amounts to nearly two-thirds of all the plastic waste, as shown in Figure 7.6. [13] The difference in the demand for packaging (34%) and the generation of packaging waste (63%) is based on the different service life of the products. About 60% of all plastic products were designed for a long service life (years), while about 40% were designed for a shorter service life or even for a single use.

Due to their similar density, packaging wastes, especially LDPE, HDPE, and PP, are difficult to separate (Section 3.3.2). Unfortunately, waste management strategies are not developing at the same rate as the increasing levels of plastic waste. In 2017, only 3 million tons (9.23%) out of the total 32.5 million tons of plastic waste in America's MSW stream were recycled (see Table 2.1). [11, 14] The recovered plastics were mainly PET and HDPE bottles. The main polymers used for packaging applications are summarized in Table 3.2. [13]

Table 3.2 Most Common Polymers Used in Packaging Applications

Application	Most Common Polymers Used
Bottles, flasks	PET (66%), HDPE (28%), PP (3%), LDPE (0.4%), PVC (3%)
Closure items, bottle caps	PP (73%), HDPE (20%), LDPE (5%), PVC (2%)
Films	LDPE (76%), PP (20%), PVC (3%), PET (1%)
Bags, sacks	LDPE (61%), HDPE (31%), PP (8%)
Jars, boxes, tubs	PP (73%), HDPE (20%), LDPE (5%), PVC (2%)

Abbreviations: HDPE, high-density polyethylene; LDPE, low-density polyethylene; PET, polyethylene terephthalate; PP, polypropylene; PVC, polyvinyl chloride

3.2.2.2 Building and Construction Plastic Waste

In the building and construction industry, plastics play an important role due to their durability, aesthetics, easy handling, and high performance. Normally, they are designed to be durable for 30 to 40 years before disposal. They are used, for example, in pipework, insulation, wall coverings and flooring, interior fittings and window frames. Common plastics used in construction include HDPE, polyvinyl chloride (PVC), and polyurethane (PU) (see Table 3.3). [13]

Table 3.3 Most Common Polymers Used in Building and Construction Applications

Application	Most Common Polymers Used
Pipes and ducts	PVC, PP, HDPE, LDPE, ABS
Insulation	PU, EPS, XPS
Windows and other frames, flooring, and wall coverings	PVC
Lining	PE, PVC
Interior fittings	PS, PMMA, PC, POM, PA

Abbreviations: ABS, acrylonitrile butadiene styrene; EPS, expanded polystyrene; HDPE, high-density polyethylene; LDPE, low-density polyethylene; PA, polyamide; PC, polycarbonate; PE, polyethylene; PMMA, polymethyl methacrylate; POM, polyoxymethylene; PP, polypropylene; PS, polystyrene; PU, polyurethane; PVC, polyvinyl chloride; XPS, extruded polystyrene

Rigid PU foam is known for its high thermal resistance, which promotes temperature maintenance. It is also popular because it is lightweight, chemically resistant, and flame retardant. Moreover, acrylonitrile butadiene styrene (ABS) and polycarbonate (PC) are used in the construction industry as well.

The variety of material grades and properties used in the building and construction industry leads to a small recycling rate for these materials. [15] Nevertheless, the building and construction industry can be counted as a secondary market for recycled plastic. There it can be used in many applications including as a filler, in packaging, in landscaping for walkways, bridges, fences, and signs, and in traffic management and industrial strapping products. [16]

3.2.2.3 Automotive Plastic Waste

In automotive applications, plastics are used for a wide variety of parts and functions, mainly because of their low weight and cost. Furthermore, they have advantages due to their impact and corrosion resistance. The largest share of plastics used in vehicles is for the passenger cell, followed by the body. PP, PE, PU, and PVC are the most common plastics by volume used in a typical car. Table 3.4 lists the common plastics used in various automotive parts.

Table 3.4 Most Common Polymers Used in Automotive Applications

Application	Most Common Polymers Used
Bumper	PP, ABS, PC/PBT
Seats	PU, PP, PVC, ABS, PA
Dashboard	PP, ABS, PPE, PC
Fuel system	HDPE, POM, PA, PP, PBT
Body	PP, PPE
Interior trim	ABS, PP, PBT, POM, PP
Lighting	PC, PBT, ABS, PMMA

Abbreviations: ABS, acrylonitrile butadiene styrene; HDPE, high-density polyethylene; PA, polyamide; PBT, polybutylene terephthalate; PC, polycarbonate; PMMA, polymethyl methacrylate; POM, polyoxymethylene; PP, polypropylene; PPE, polyphenylene ether; PU, polyurethane; PVC, polyvinyl chloride

3.2.2.4 Agricultural Plastic Waste

Plastics are everywhere on a typical farm and substitute for traditional materials due to their low price and light weight. Nowadays, hay bales are often wrapped in plastic or grain is stored in plastic bags instead of silos. Furthermore, plastic has become essential for widespread applications in modern farm operations. For example, large plastic sheets are used for everything from heating the soil and suppressing weeds to roofing greenhouses. A lot of agricultural products are plastic films, which is reflected in the types of polymers used in agricultural applications as shown in Table 3.5. Currently only 10% of agricultural plastic waste is recycled. [13, 17, 18, 19]

Table 3.5 Most Common Polymers Used in Agricultural Applications

Application	Most Common Polymers Used
Bale bags, seed bags	LDPE, LLDPE, PP
Greenhouse covers, silo covers, mulch film	LDPE, LLDPE
Nets and mesh	LDPE, HDPE
Rope, strings	PP
Pipes and fittings	PVC, LDPE
Pesticide containers, nursery pots	HDPE, PS, PP

Abbreviations: HDPE, high-density polyethylene; LDPE, low-density polyethylene; LLDPE, linear low-density polyethylene; PP, polypropylene; PS, polystyrene; PVC, polyvinyl chloride

3.2.2.5 Waste from Electrical and Electronic Equipment (WEEE)

In general, the plastic waste from electrical and electronic equipment (WEEE) and in particular the metals in them are recovered by metal recyclers. The remaining plastics and nonmetals are known as electronics shredder residue (ESR). The principal polymers found in WEEE include ABS, high-impact polystyrene (HIPS), PP, polyamide (PA), polycarbonate (PC), blends of PC with ABS (PC/ABS), blends of polyphenylene ester (PPE) and HIPS (PPE/HIPS), and some others as shown in Table 3.6. So far only half of a typical ESR mixture is recovered and the rest goes into landfill. [13, 20, 52]

Table 3.6 Most Common Polymers Used in Electrical and Electronics Equipment

Application	Most Common Polymers Used
Printers/faxes	PS, HIPS, SAN, ABS, PP
Telecommunications equipment	ABS, PC/ABS, HIPS, POM
Televisions	PPE/HIPS, PC/ABS, PET
Monitors	PC/ABS, ABS, HIPS
Computers	ABS, PC/ABS, HIPS
Refrigeration	PS, ABS, PU, PVC
Dishwashers	PP, PS, ABS, PVC

Abbreviations: ABS, acrylonitrile butadiene styrene; HIPS, high-impact polystyrene; PC, polycarbonate; PET, polyethylene terephthalate; POM, polyoxymethylene; PP, polypropylene; PPE, polyphenylene ether; PS, polystyrene; PU, polyurethane; PVC, polyvinyl chloride

3.3 Sorting Processes for Plastic Waste

After postconsumer plastic waste is collected, it is transported to material recovery factories (MRF) and, in a first step, sorted by plastic type. Depending on the type of plastic, sorting methods vary. Sorting currently is the step in the recycling process of postconsumer waste that has the largest impact on integrity. Sorting can be done manually or it can be automated. In the following discussion, the main methods used in the MRF in the United States for the different plastics are identified.

3.3.1 Manual Sorting

A simple sorting method is *manual sorting*. It involves visual identification of the plastic type by operators using the resin identification code, shape, color, appearance, and trademark of the plastic. It is very labor intensive and has a possibility of human error. Furthermore, it is difficult to manually differentiate between the resin types by just visual means.

3.3.2 Automated Sorting

3.3.2.1 Float-and-Sink Sorting

The *float-and-sink* process, or sorting by flotation, in which plastics are sorted by density is one of the most common *automated sorting* processes. The washed and chipped plastics are sent into tubs of water and the pieces that float or sink are separated. This process is fast, inexpensive, and can be considered as a first stage washing of plastic waste. However, as pointed out earlier, most plastics are very similar in density and thus cannot be separated using this process. An overview of the densities of various polymers is given in Table 3.7. [9, 21, 22]

Table 3.7 Polymer Density Ranges (densities of < 1 g/cm^3 will float)

Polymer	Density Range [g/cm^3]
Polyethylene terephthalate (PET)	1.330–1.400
High-density polyethylene (HDPE)	0.956–0.980
Polyvinyl chloride (PVC)	1.304–1.388
Low-density polyethylene (LDPE)	0.910–0.955
Polypropylene (PP)	0.861–0.925
Polystyrene (PS)	1.050–1.220

3.3.2.2 Froth-Flotation Sorting

The *froth-flotation* process works similarly to the float-and-sink process. In froth-flotation, the materials to be separated are first treated with a surfactant and then suspended in water. Plastics that would normally sink in water are suspended in the water-surfactant mixture. Then, air is pumped into the system. The air bubbles adhere to some plastic pieces based on their resin type but others are not affected by the air bubbles and so sink to the bottom. The key advantage of this method is to be able to separate PET from PVC.

3.3.2.3 Near-Infrared Sorting

Another promising technology suitable for high-speed sorting machines is *near-infrared (NIR) sorting,* which uses a well-established technique for characterizing the type of plastic, infrared light transmission. Plastics absorb light of specific wavelengths unique to their chemical composition, allowing for the identification and separation of different types of plastic.

This technology is used for the automated identification of PET bottles. However, this sorting method is not suitable for identification of dark-colored plastics, adhesives, residues, and plastics with additives. [23, 24]

3.3.2.4 Laser-Aided Identification

Laser-aided systems identify plastics by shining a laser beam onto the material surface to determine the material's spectral and spatial characteristics for identification and location. This is also used on high-speed sorting lines.

Other laser based systems can determine various material properties including absorption coefficient, thermal conductivity, thermal capacity, and surface temperature distribution, which can be used to identify the plastic type.

3.3.2.5 X-Ray Fluorescence

Some plastics, like PET and PVC, have similar densities, and so sorting them must be based on another property in which they differ significantly. The molecular structure of PVC includes chlorine atoms, unlike PET, which does not. *X-ray fluorescence* generates a spectral fingerprint based on a plastic's chemical composition, and, from this, it can be sorted into its resin type. For example, due to the detection of the chlorine atoms in PVC, it can be used to separate PVC from other plastics. [25]

3.3.2.6 Marker Systems

Marker systems involve marking either the plastic container or the resin itself with something readily detectable. One marker system, developed by Continental Container Corporation, allows for sorting by resin type, color, resin additive package, and package contents. This method is costly since every packaging production line would need to install a marking system and the recycler would need to install a machine for scanning the marker. An alternative method developed by the Eastman Chemical Company uses a molecular marker to identify the resin type. The cost increase for the addition of the marker system would be relatively low since it would only require recyclers and resin manufacturers to contribute to the costs. However, all of these methods are still at the proof-of-concept stage in development.

3.4 Plastic Degradation Mechanisms

In general, polymers are stabilized only for processing and a first lifetime of use and not for reprocessing or further lifetimes of use. This is particularly true for plastic film packaging material made out of polyolefins, such as PP and PE, which usually has a short lifetime.

Degradation of a polymer results from exposure to factors like heat, mechanical stress, oxygen, and UV light. For example, polyethylene films used in commercial packaging suffer from oxidative and photo-oxidative degradation due both to processing and to use. Reprocessing can induce mechanical, thermal, and thermal oxidative degradation.

The degradation process leads to changes in the polymer's structure, typically characterized by fragmentation, or scission, of the macromolecular chain, which results in changes in the properties of the polymer. Radical chain reactions, such as formation of hydroperoxides and cross-linking, also occur during degradation and also result in changes in the polymer's structure. During processing, both chain scission and cross-linking reactions take place, which lead to increased fractions of lower molecular weight polymer chains and increased chain branching, respectively. Thus, fragmentation, self-termination, and chain-transfer reactions of the polymer radicals are occurring during degradation. [26] Polyolefins often undergo oxidative degradation.

During processing and lifetime use, the material is exposed to heat, mechanical stresses, oxygen, and UV radiation. Depending on the polymer structure and the processing conditions, these stresses lead to more or less degradation. In general, the effects of the degradation processes can be categorized as:

- Change in molecular weight and molecular weight distribution, leading to a change in viscosity
- Formation of cross-links and branched chains
- Formation of oxygenated compounds and unsaturated compounds

A knowledge of the degradation kinetics of polymers is important for designing the recycling process and the processes for creating new products out of recycled material. During the recycling process, the *melt viscosity* and *flow behavior* of materials change, leading to a change in the processing settings and sometimes difficulty in reprocessing. The recycled products can further have inferior mechanical properties compared to virgin material. The thermal stability is not affected for most polymers.

The mechanism of degradation, which can be classified as *mechanical degradation, thermal degradation, thermal oxidative degradation*, or *photo degradation*, depends on the method used for polymerization. Furthermore, the degree and type of degradation depend on the processing conditions and on the nature of the polymer.

3.4.1 Mechanical Degradation

The chemical bonds of the carbon-carbon polymer backbone break when shear and tensile stresses (mechanical stresses), induced by shearing and stretching in the extrusion process, exceed the intramolecular bonding forces leading to mechanical degradation. The mechanical stresses increase with increasing chain length and decreasing distance to the center of the chain. As a result, chain scission increases for longer polymer chains and is most likely to occur in the chain center. In addition, a reduction in temperature decreases the flexibility of the chain segments, which also leads to an increase of these mechanical stresses. [27, 28]

3.4.2 Thermal Degradation

Thermal degradation is induced by heat. Heating of a polymer results in an increase of its internal energy causing the following effects: First, the rate of rotation of any freely moving group in the polymer increases, weakening the intermolecular forces. Second, the vibrational energy of the polymer bonds increases, leading to bond breaking along the chain according to a statistical pattern and the formation of two radicals for each end of the broken bond. Finally, the mobility of absorbed species increases, enabling their migration through the polymer and hence their reaction with energetic sites. [29, 30]

3.4.3 Thermal Oxidative Degradation

Thermal oxidative degradation of polymers is caused by autoxidation. The reaction follows the same steps known for polymerization: initiation, propagation, and termination. Depending on the dominating reactions during autoxidation, both a decrease or an increase of the average molecular weight of the polymer can occur. While β-scission and fragmentation cause a decrease of the molecular weight, an increase can be observed for recombination. [31, 32, 33]

3.4.4 Effect of Degradation on Processing and Service-Life Properties

It is important to analyze the effect of degradation on the processing parameters as well as the service-life properties of the desired product. Based on the product requirements, the number of reprocessing events has to be limited or the amount of virgin material mixed into the recyclate has to be adjusted to reduce the dilution effect. In the following, data for these property changes is presented for some common unfilled and filled plastics.

3.4.4.1 Unfilled Plastics

As already mentioned, PET is the most widely recycled polymer due to the ease of its separation in the recycling stream. The degradation mechanism of PET is chain scission, which can be seen from the reduction of molecular weight from virgin PET to recycled PET over three extrusion cycles (Figure 3.3 *left*). Even though the chain length is reduced significantly, the mechanical properties are only slightly affected as can be seen in Figure 3.3 *middle* and *right*. Both tensile strength and impact strength show a small reduction over the three extrusion cycles.

In industrial practice, virgin PET is often used to refresh the recyclate. The mixing ratio and resulting dilution (Figure 3.1) depends on the desired product properties. Figure 3.4 shows the change of the molecular weight and impact strength for different mixing ratios. The graph in Figure 3.3 *left* shows that the molecular weight of 100% recycled PET (same as reprocessing step 2 in Figure 3.3) is about 20% lower than virgin PET (same as reprocessing step 1 in Figure 3.3). The blended materials follow the linear trend based on the mixing ratio of virgin to recycled PET. Both data sets show the feasibility of reprocessing and the ability of refreshing with virgin material to slow down the degradation process significantly.

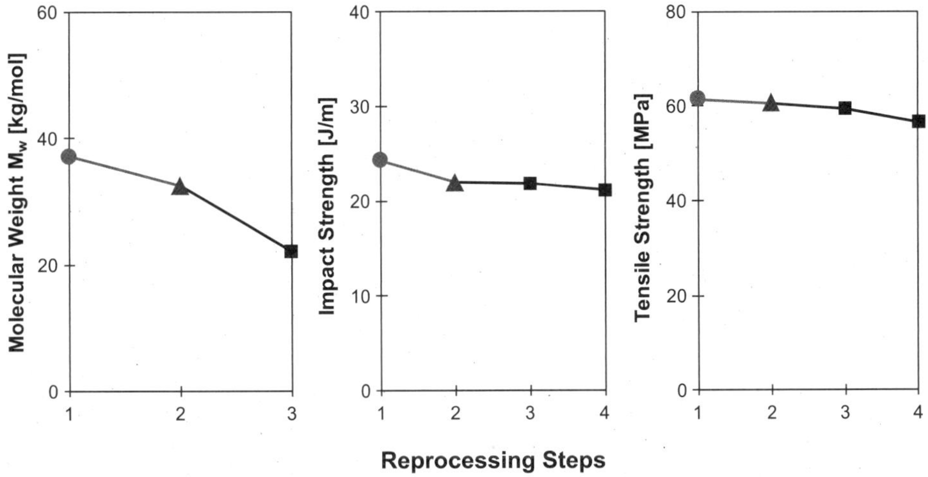

Figure 3.3 Weight-averaged molecular weight and mechanical properties of virgin polyethylene terephthalate (PET) and recycled PET as a function of the number of reprocessing steps (extrusion) [34]

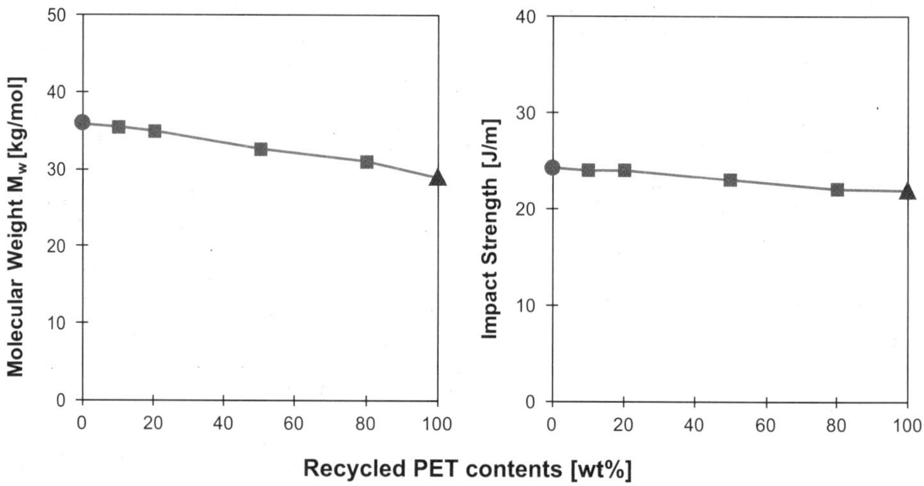

Figure 3.4 Weight-averaged molecular weight and impact strength of virgin PET (polyethylene terephthalate)/recycled PET blends during the first extrusion step [34]

A decrease in molecular weight leads to a reduction in melt viscosity, which is a measure of the resistance to flow, and thus affects the processing behavior. Therefore, the process settings have to be adjusted accordingly. For example, the injection pressure can be lower. However, if the mixing ratio is kept constant, the material is in a steady state and adjustments are very minimal. Nevertheless, in certain cases the processibility of materials is affected so strongly that the material cannot be processed with the same process. The stretch-blow molding process is an example of such a sensitive process.

High-impact polystyrene (HIPS) is among the promising materials for recycling since it shows only small variations in melt viscosity. Figure 3.5 shows the change in viscosity measured as melt flow index (MFI), which indicates how much material in grams flows through a small capillary within 10 min. The larger the MFI values, the lower the viscosity. The MFI increases slightly with an increasing number of reprocessing steps. The increase in MFI during this continued reprocessing without refreshing is caused by chain scission degradation due to thermal and mechanical degradation.

Figure 3.6 shows the discoloration of HIPS due to degradation during the various reprocessing steps. Furthermore, it illustrates the resulting changes in the tensile properties of HIPS. The degradation leads to an increase of tensile stress at break and a reduction of elongation at break over multiple reprocessing steps. Thus, the material behavior changes from more ductile to brittle, which needs to be known for product design. [35, 36]

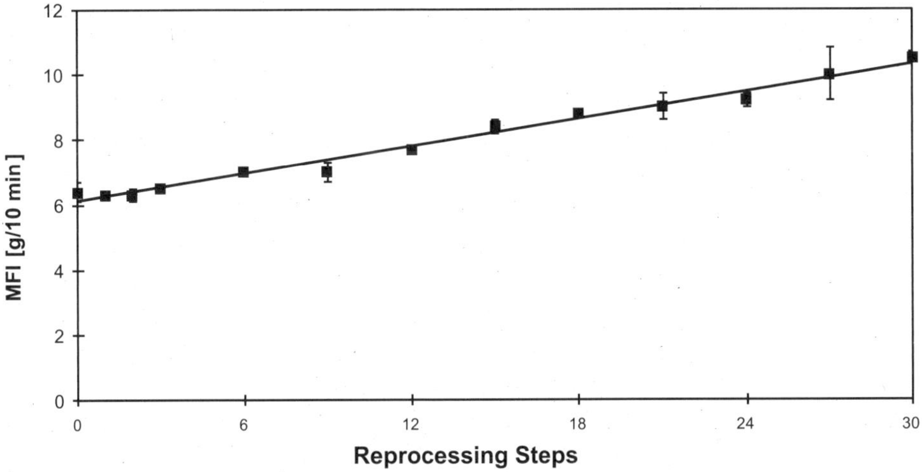

Figure 3.5 Melt flow index (MFI) of high-impact polystyrene (HIPS) as a function of the number of reprocessing steps (injection molding) [35]

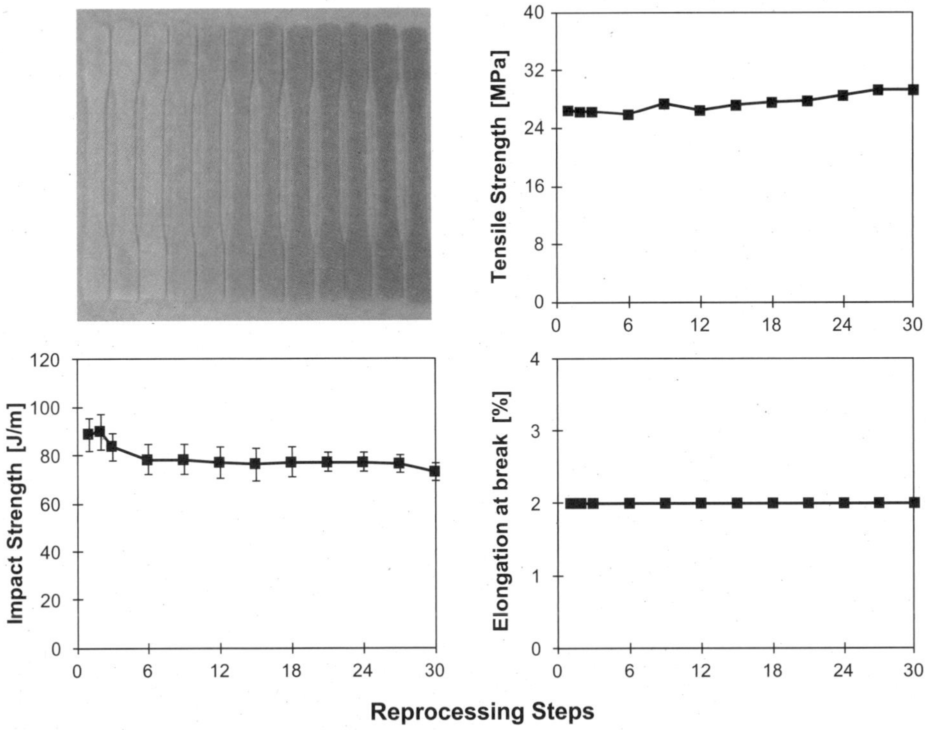

Figure 3.6 Discoloration and mechanical properties of high-impact polystyrene (HIPS) as a function of the number of reprocessing steps (injection molding) [35]

Instead of chain scission, degradation can lead to uncontrolled cross-linking of polymer chains for some polymers. This can be observed in the data of LDPE obtained over 100 extrusion cycles (Figure 3.7). Here, the complex viscosity of LDPE *(left)* increased with the increasing number of extrusion cycles. The cross-linking in LDPE chains occurs due to the formation and reaction of carbon radicals. The same trend was observed in the decreasing MFI values *(right)*. It is important to note that the reprocessing of LDPE is only significantly affected after the 40th extrusion cycle. Using a refreshing rate of less than 10% of virgin material results in a fraction of < 1% of 40 times processed material according to Equation 3.1 and thus has no practical relevance in this case. [5]

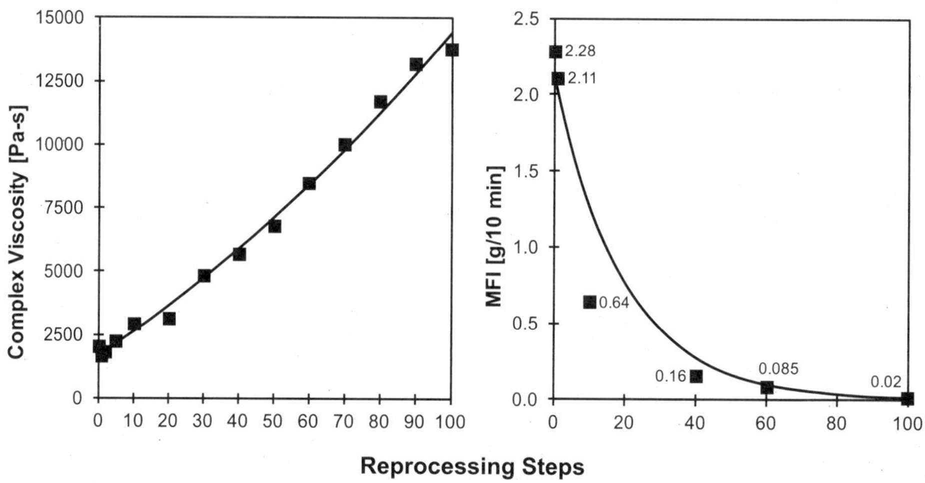

Figure 3.7 Complex viscosity and melt flow index (MFI) of low-density polyethylene (LDPE) during 100 reprocessing steps (extrusion) [5]

In addition to synthetic polymers, more and more biopolymers, which are made from renewable resources, are being recycled. One major market for biopolymers is plastic packaging. Polylactic acid (PLA), which is biodegradable under very specific conditions, is such a polymer. Since the conditions for chemical recycling can only be achieved in specific recycling facilities, the mechanical recycling of PLA is of interest as well. Figure 3.8 shows that multiple injection molding cycles cause degradation, which can be seen in the significant change in MFI. However, the impact strength of PLA remained unchanged even after six cycles, which indicates good recyclability in regards to the service-life properties of the plastic.

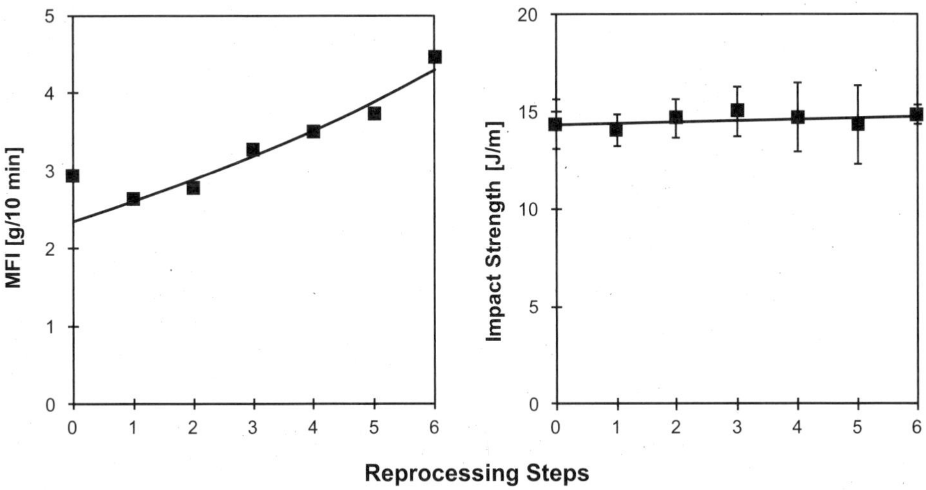

Figure 3.8 Melt flow index (MFI) and impact strength of polylactic acid (PLA) as a function of the number of reprocessing steps (injection molding) [37]

Polycarbonate (PC) is another material with a high potential for recycling. It is identified as plastic number 7 and mostly found in electronic waste. Recycling of PC is important to prevent the leaching of bisphenol A in PC into landfills. Figure 3.9 illustrates the effect of repeated injection molding cycles on the mechanical properties of PC. Tensile strength and modulus remain unchanged after five cycles. In contrast, the elongation at break and the toughness decrease with the increasing number of reprocessing cycles. The toughness of PC shows an approximately 42% reduction from the virgin resin to the final reprocessing steps, whereas the elongation at break shows a 37% reduction. [38]

Again, comparing the property changes after multiple processing steps of 100% of the material (Figure 3.9) with the change in refreshed material reveals almost unchanged behavior during refreshing. Figure 3.10 shows the same mechanical properties, but for different mixing ratios of virgin and recycled PC. It can be seen that the material follows the linear trend depending on the mixing ratio, similar to the observations with PET.

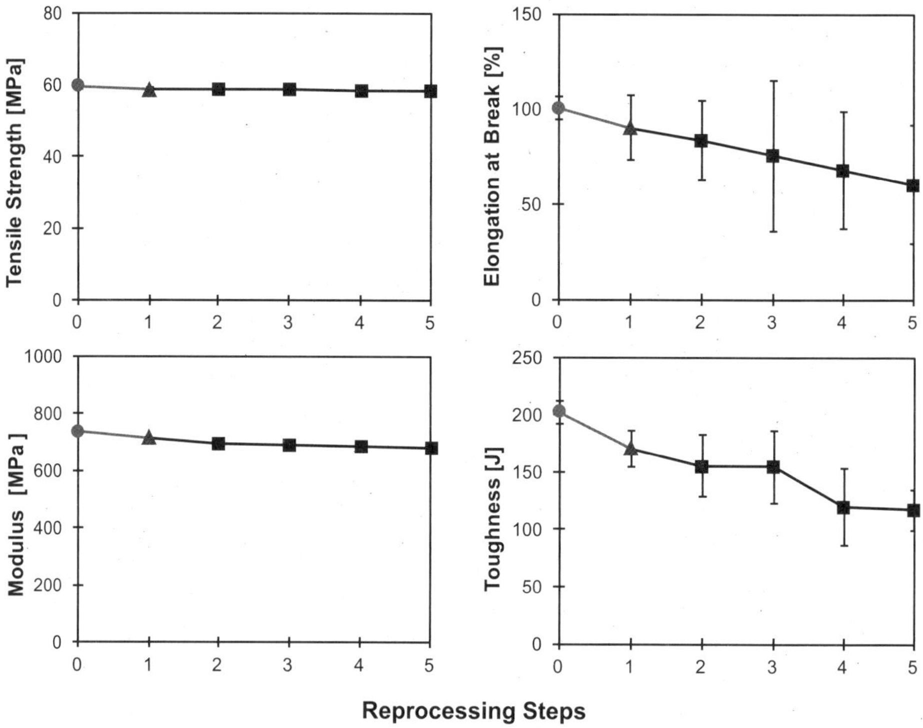

Figure 3.9 Mechanical properties of polycarbonate (PC) as a function of the number of reprocessing steps (injection molding) [38]

Stabilizers used for recycled plastics are mostly phenolic antioxidants, phosphites, or phosphonites combined with co-stabilizers, hindered amines, sterically hindered piperidines, and UV absorbers. Pfaendner [53] and Pospisil et al. [54] provide a good overview. They show in one study that a mixture of two additives, one used to improve the long-term stability of PP and the other a UV stabilizer with additional enhancement capability for processing and long-term stability of HDPE, significantly increases the time to embrittlement of filled PP; see Figure 3.11. [53] The former garden chairs can now even be used for the same application. The embrittlement can be reduced further when additional antioxidants are added. This shows the huge potential of stabilizers, but also the complexity of selection and combination of these additives.

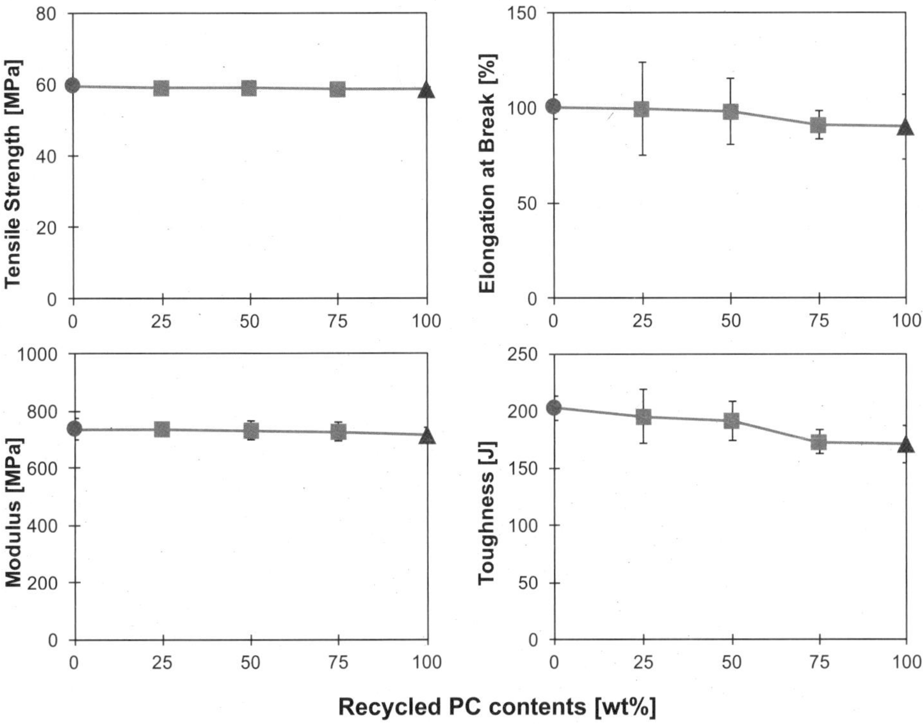

Figure 3.10 Mechanical properties of virgin polycarbonate (PC)/recycled PC blends as a function of the percentage of recycled PC, in the first reprocessing step [38]

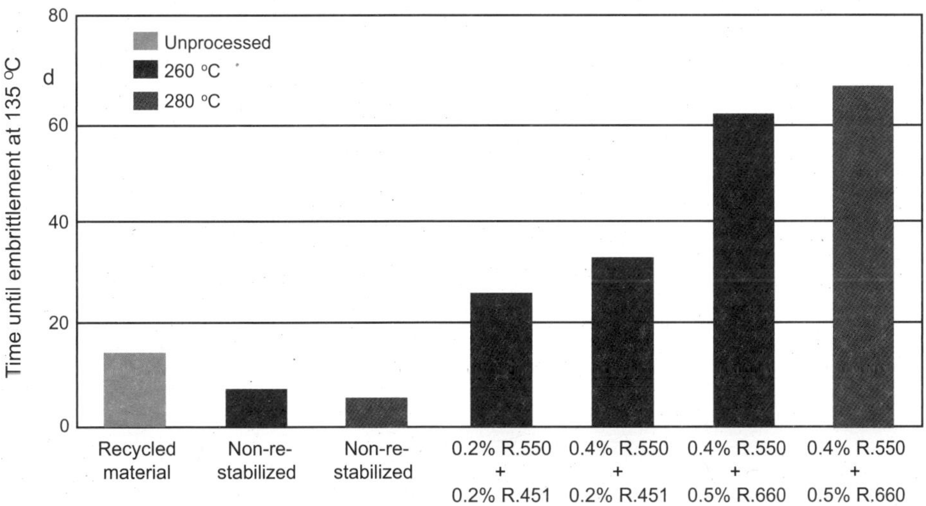

Figure 3.11 Aging stability of a $CaCO_3$-filled, recycled PP material from garden chairs at different processing temperatures (R. 451 = Recyclostab 451, R. 550 = Recyclossorb 550, R. 660 = Recycloblend 660) [53]

The mechanical recycling of postconsumer waste leads inevitably to the formation of polymer blends and most common polymers are immiscible in the melt phase. This usually results in lower mechanical performance, often referred to as downcycling. The reason is phase separation and low adhesion between the two or more phases, which leads most commonly to low impact strength. The solution for virgin blends can be applied to blends of recycled plastics as well: compatibilizers.

Compatibilizers for blends from recycled plastics are available for polyolefin blends of PP and PEs, polyolefins contaminated with polar polymer (e. g., PA) from plastic films, and PET or PA contaminated with low amounts of PEs or PP (from bottle caps). Figure 3.12 shows the effect of the compatibilizer SEBS-g-MAH on the otherwise immiscible blend of PP/PET. Both the elongation at break and the impact strength increase with the additive content, highlighting the ductile behavior of the now miscible blend. [53]

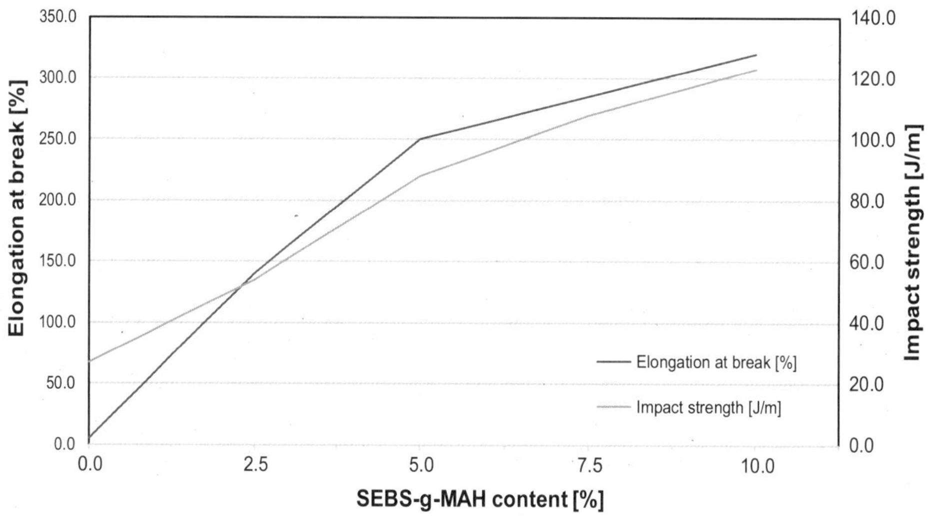

Figure 3.12 Effect of compatibilizers as illustrated by a PP/PET blend [53]

3.4.4.2 Fiber-Reinforced Plastics

Dr. Sebastian Goris

Fiber-reinforced plastics (FRP) are a distinctive class of materials defined by the combination of at least two constituents: the matrix material and the reinforcing fibers. By merging the separate phases, the composite has enhanced properties that exceed the performance and capabilities of the individual constituents. Fiber-reinforced composites are classified into *continuous* and *discontinuous* fiber-reinforced composites. Fiber-reinforced composites are commonly applied as a substitute for metal as they provide key advantages in form of lower manufacturing costs, component integration, lower density, and high-volume production.

Continuous fibers for reinforcement can be used in many forms, such as woven, fabrics, knitted constructions, or laminated composites. Laminated composites consist of multiple layers of unidirectional fibers each oriented differently, to optimize performance at laminate level. Laminated composite materials are commonly used for high-performance applications, because of their superior mechanical properties.

Discontinuous fibers are another form factor of the fibers within the composite in which fibers are used to improve the performance of the base resin. This material class is characterized by the fact that the fibers are chopped and experience a change in orientation and a reduction in length during molding. This process-induced change in the fiber microstructure determines the performance of the molded part.

FRP materials are used in various industries, such as automotive, aerospace, construction, consumer goods, sporting goods, oil and gas, and wind power. Their utilization has seen continuous growth, particularly for applications that replace metals for lightweight design. The recent increase in production due to the many applications will inevitably result in an increase in the amount of end-of-life material and waste. Hence, the question of recycling for FRP has become a major topic of research in academia and industry. While most challenges in the recycling of unfilled plastics also apply to FRP, there are additional considerations for recycling fiber-reinforced materials. In particular, the reduction of the length of the fibers through additional reprocessing can have a profound impact on the value of recycled FRP parts.

3.4.4.2.1 Discontinuous Fiber-Reinforced Composites

Discontinuous fiber-reinforced composites comprise either a *thermoset* or a *thermoplastic* matrix. Sheet molding compound (SMC) is the most prominent type of discontinuous fiber-reinforced thermoset and is processed in compression molding. [41] A variety of thermoplastic matrices and processes are available for discontinuous fiber-reinforced composites, while engineering thermoplastics, such as polyamide (PA) or polypropylene (PP), represent the majority of matrices. [42] Glass fibers are frequently used for reinforcement due to their availability, low cost, and high strength. Although carbon fibers can offer superior performance compared to glass fibers, the cost increase often does not justify their use as substitute in FRP for all applications. [43] Nevertheless, recycling of carbon fibers has become an active field of research aiming to use scrap fiber material from continuous carbon fiber processes and reuse it in discontinuous fiber-reinforced compounds. [44]

Discontinuous fiber-reinforced thermoplastics can be further classified as *short fiber-reinforced thermoplastics* (SFT) and *long fiber-reinforced thermoplastics* (LFT). Both classes are delivered in pellet form, as schematically depicted in Figure 3.13. However, the two types are differentiated in the initial length of the fibers within the pellet before processing. Another way to look at the distinction is the average fiber length to fiber diameter ratio, also called the *fiber aspect ratio*. If the ratio is below 100, the composite is commonly classified as an SFT. If the value exceeds 100, it is referred to as an LFT. The performance and cost of LFT compounds usually place them between continuous fiber-reinforced composites used for high-performance applications and SFC compounds.

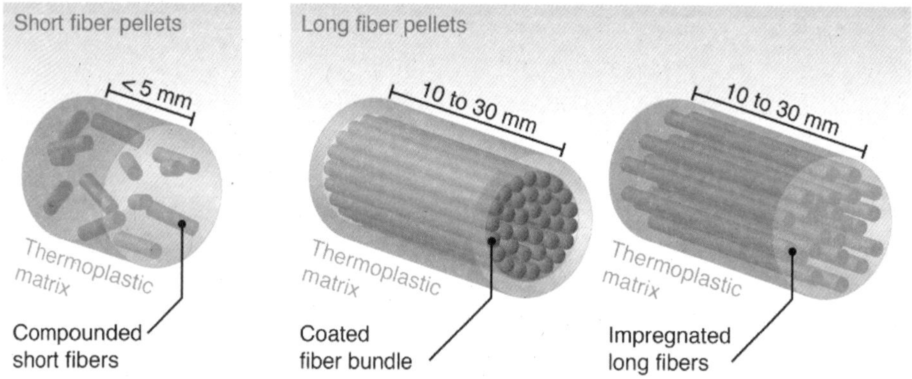

Figure 3.13 Illustration of short fiber pellets (left) and long fiber pellets, which are further divided in coated fiber bundles (center) and pultruded fibers (right) [41]

3.4.4.2.2 Fiber Length Reduction

The reinforcing characteristic of the fibers is achieved by transferring the applied load from the matrix to the fibers. Consequently, the interfacial adhesion as well as the length of the fibers embedded in the matrix determine the maximum transferable load and the overall reinforcement. The effect of the residual length of the fibers on the mechanical properties of a molded part is illustrated in Figure 3.14. The graph shows the effect of the fiber aspect ratio on the tensile modulus, tensile strength, and impact strength. [45, 46, 47, 48] The figure also qualitatively illustrates the expected range of the residual fiber aspect ratios in LFT injection molding, as reported in the literature. [41]

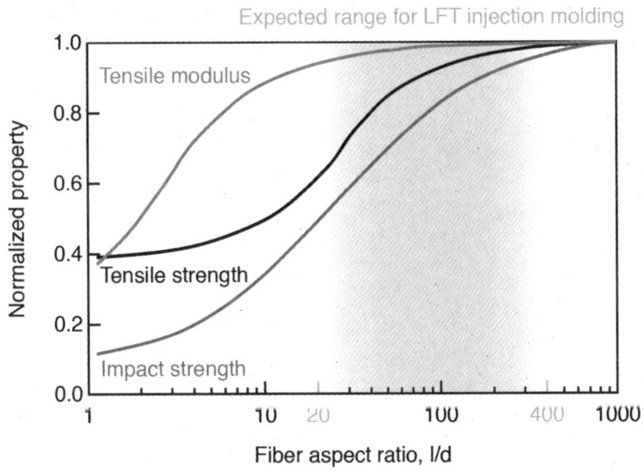

Figure 3.14 Normalized mechanical properties as a function of fiber aspect ratio for tensile modulus, tensile strength, and impact strength. The graph is adapted from Thomason et al. [45, 46, 47, 48] The highlighted range illustrates the fiber aspect expected in LFT injection molding

Preserving a fiber length that is sufficiently long to transfer the maximum load, while maintaining economical process cycles, is a major challenge for molding SFT and LFT materials. While the initial length of the fibers may be uniform and up to 30 mm for an LFT compound, fiber attrition results in a distribution of fiber lengths in the finished part after only one processing event ranging from powder-like fibers (< 0.1 mm) to a small fraction of fibers retaining their initial length. In general, the degree of fiber attrition is influenced by the process, processing parameters, material selection, and mold design. [55] On the micro-scale, the driving mechanisms for fiber attrition are hydrodynamic effects, fiber–fiber interactions, and fiber–equipment interactions, which the fibers experience along all stages of the process. For a comprehensive review of the underlying physics of fiber attrition during processing, the reader is referred to [49] and [41].

Lafranche et al. [50] conducted a comprehensive experimental study to evaluate the fiber length reduction along all stages of a single injection molding cycle. Figure 3.15 shows the measured fiber length from the pellets (L_N = 7.9 mm) to the molded part (L_N = 1.9 mm), which is an overall fiber length reduction of more than 75%. It can be seen that the majority of the length reduction happens in the screw sections and only small changes occur in the runner and cavity. Furthermore, this shows that the improvement in mechanical performance of injection molded FRP can be mainly attributed to a fiber length much shorter than in the pellet.

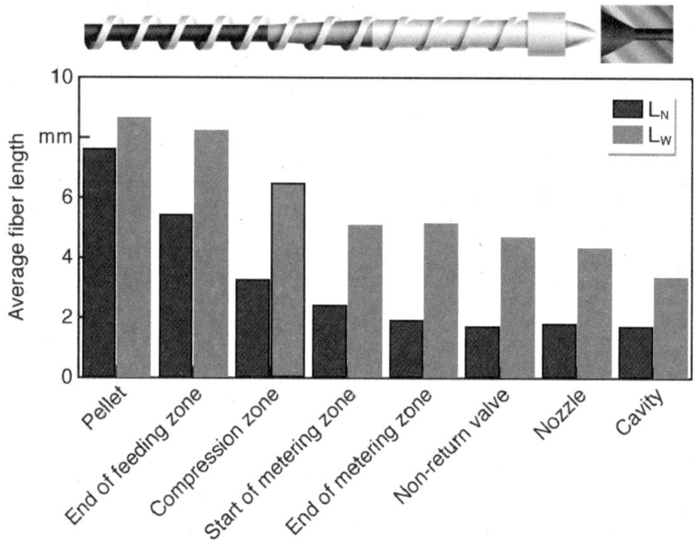

Figure 3.15 Fiber breakage during the plasticizing stage, adapted from [50]. L_N is the number-average fiber length and L_W is the weight-average fiber length. While L_N represents the mean of the measured fiber population, the characteristic of L_W is a weighted average emphasizing the proportion of long fibers in the distribution

Further studies indicate that the residual fiber length varies substantially within a molded part. [41, 49] Depending on the material properties, fiber concentration, processing conditions, and mold geometry, the local fiber length within the molded part can vary up to 30%. [51] This finding further highlights the heterogeneous fiber microstructure seen in discontinuous fiber-reinforced thermoplastics and it shows that the common assumption of a uniform fiber length throughout a molded part is not valid. Nevertheless, the majority of the fiber length reduction occurs during the plastification stage.

A study of fiber length reduction under controlled conditions using steady shear in a customized Couette rheometer showed the continuous reduction of fiber length in processing LFT compounds. [49] Polypropylene at 30 wt % fiber loading was exposed to steady shear conditions under isothermal melt temperature (250 °C). As shown in Figure 3.16, the fiber length rapidly decreases from 15 mm down to approximately 1.6 mm for a speed of 50 RPM or 1 mm for a speed of 150 RPM. Even at a low residence time (< 60 seconds), the residual fiber length is only approximately 27% of the initial fiber length. This highlights the sensitivity of fiber attrition and processing, which consequently poses a major challenge in recycling fiber-reinforced composites. From this experimental study, it can be inferred that an *unbreakable fiber length* exists, at which point no additional fiber breakage occurs for a given processing condition. [49] While the correlation between the idealized conditions in the Couette rheometer and conventional molding processes is still subject to research efforts, [49] the work suggests that there is a lower bound on residual fiber length that is reached in recycling of discontinuous fiber-reinforced plastics.

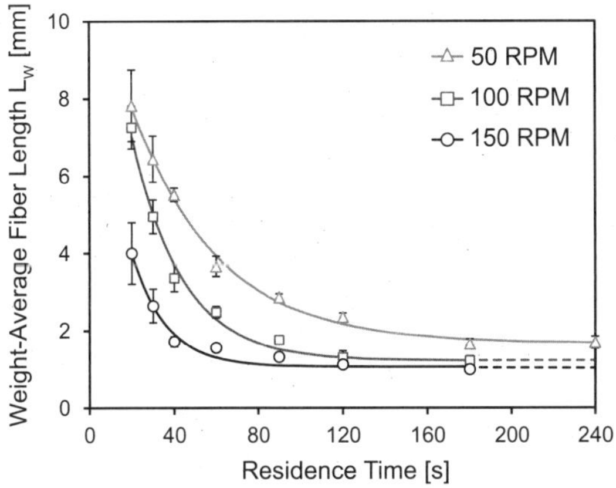

Figure 3.16 Fiber length reduction for glass fiber-reinforced polypropylene under steady shear, adapted from [49]. A 30 wt % filled polypropylene was processed under isothermal melt temperature in a customized Couette rheometer. The residual fiber length was measured at varying residence times to understand the evolution of fiber attrition

3.4.4.2.3 Recycling of Fiber-Reinforced Composites

The premium paid for the reinforcing fibers over unfilled resins, particularly for more expensive fibers such as carbon fibers, warrants the consideration of recycling the fiber-reinforced composites. While the application of this material class is increasing, recycling and end-of-life considerations remain a mostly academic research topic. Major challenges in recycling remain, mostly attributed to the fact that fiber-reinforced composites are made from a composition of several constituents with differing material characteristics. For the most simplistic fiber-reinforced composite, the constituents include at least the fibers, the matrix, and the fiber sizing. For now, complete separation and recycling of the individual components is challenging, if not impossible, in terms of economic considerations.

Additionally, fiber length reduction is a major challenge since the average fiber length is a key determinant of the mechanical performance of the molded part. Any reprocessing of fiber-filled compounds can result in additional reduction of fiber length and, consequently, lower the mechanical performance. Studies [39, 40] show that the recycling of manufacturing scrap leads to significant deterioration in the mechanical properties compared to the virgin material, which has been attributed to the additional reduction in fiber length, as shown in Figure 3.17. While the change in properties is substantial, it can be seen that refreshing of recycled material with virgin material can offset the loss in performance significantly, as shown in Figure 3.18.

As described in the previous section, the majority of the fiber attrition occurs inside the screw during plastification and it can be assumed that a lower bound of fiber length reduction (unbreakable fiber length) exists for given processing conditions. Hence, comparing the recycling of preconsumer to postconsumer waste, no significant difference in the final fiber lengths are expected as most of the fiber length reduction occurs during plastification.

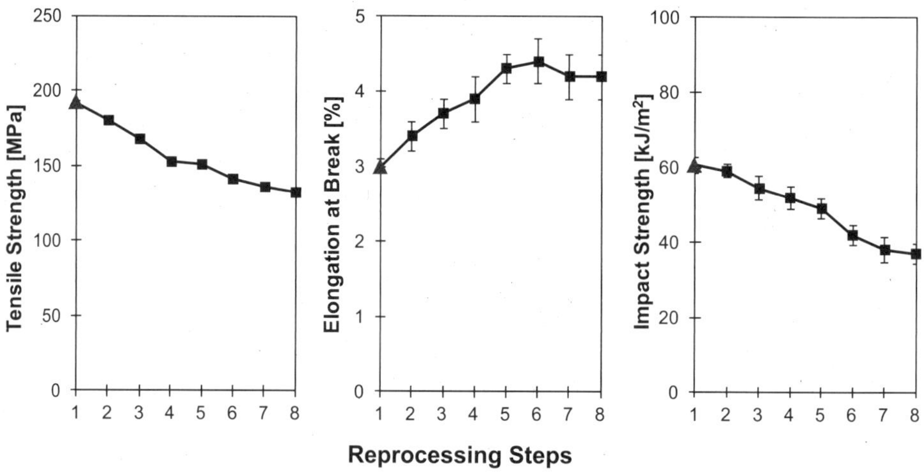

Figure 3.17 Mechanical properties of recycled fiberglass/PA66 composite as a function of the number of reprocessing steps (injection molding) [39, 40]

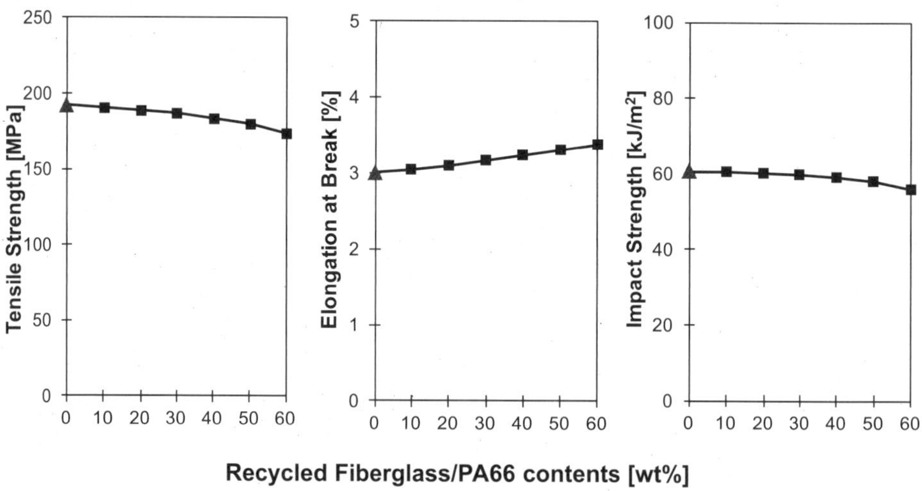

Figure 3.18 Mechanical properties of fiberglass/PA66 composite as a function of the percentage of recycled composite, in the first reprocessing step [39, 40]

3.5 Contaminants

Plastics are hard to clean due to the penetration of contaminants into the polymer matrix. Composites and mixed plastic waste are especially difficult to separate into the different plastic types, which all require different reprocessing techniques and settings. [23]

There are a number of contaminants that can significantly obstruct the recycling process and result in severe deterioration of performance of recycled material. During the extrusion process in which the recycled material is subjected to high temperatures and mechanical stresses, the presence of contaminants may lead to hydrolytic and thermal degradation and subsequent decreases in both the molecular weight and viscosity of the plastic.

In the case of the PET recycling process, the common contaminants are acids and acid-producing compounds, for example, which arise when PET and PVC are mixed. Hydrochloric acid produced from PVC acts as catalyst for chain-cleavage reactions. Likewise, elevated water content can lead to chain cleavage through hydrolysis. Most water contamination comes from the washing process and can be removed by proper drying. Dyes and coloring agents are another source of contamination, leading to undesirable, mostly brownish color in the recycled plastic. Finally, contaminants, such as acetaldehyde (a natural degradation product of PET) and other contaminants arising from misuse of PET by consumers (such as for storage of fuel, pesticides, and other dangerous materials), are potential health hazards in recycled PET products. [23]

3.6 Conclusion: Technical Feasibility of Plastics Recycling

In summary, the recyclability of plastic waste depends on the origin of the waste as well as the sensitivity of the polymer(s) to degradation. In most cases, preconsumer waste (manufacturing scrap) can be reprocessed with little deterioration of properties. The property changes can even be minimized or extended by refreshing the regrind waste with virgin plastic. Although the processing properties such as viscosity are affected, the changes in service-life properties such as mechanical performance are often negligible. In industrial production, the regrind proportion can be as high as 70%.

The recycling rate of postconsumer waste, however, is low. This is due to technical limitations such as the limited availability of clean and unmixed postconsumer plastic waste. This rate can only be increased when the recycling process becomes an integral part of the product design process and both the manufacturer and consumer take an active part in the improvement of that process. The most promising approaches for postconsumer waste to date are in chemical recycling.

References

[1] The Association of Postconsumer Plastic Recyclers and American Chemistry Council. 2013 *United States National Post-Consumer Plastics Bottle Recycling Report.* Washington, D. C. pp. 1–10, 2014.

[2] American Chemistry Council Economics & Statistics Department. *Plastic Resins in the United States.* Washington, D. C. 2013.

[3] Hopmann, C., and Michaeli, W. *Einführung in die Kunststoffverarbeitung,* 7th ed. Hanser, Munich. 2015.

[4] Ignatyev, I. A., Thielemans, W., and Vander Beke, B. Recycling of polymers: a review. *ChemSusChem.* vol. 7, pp. 1579–1593, 2014.

[5] Hamad, K., Kaseem, M., and Deri, F. Recycling of waste from polymer materials: an overview of the recent works. *Polymer Degradation and Stability.* vol. 98, pp. 2801–2812, 2013.

[6] Burillo, G., Clough, R. L., Czvikovsky, T., Guven, O., Le Moel, A., Liu, W., Singh, A., Yang, J., and Zaharescu, T. Polymer recycling: potential application of radiation technology. *Radiation Physics and Chemistry.* vol. 64, pp. 41–51, 2002.

[7] Dandy, L. O., Oliveux, G., Wood, J., Jenkins, M. J., and Leeke, G. A. Accelerated degradation of polyetheretherketone (PEEK) composite materials for recycling applications. *Polymer Degradation and Stability.* vol. 112, pp. 52–62, 2015.

[8] MainStream Project. *The New Plastics Economy: Rethinking the Future of Plastics.* Ellen MacArthur Foundation, World Economic Forum, and McKinsey & Company. Geneva, Switzerland. 2016.

[9] Goodship, V. *Introduction to Plastics Recycling.* Smithers Rapra, Shawbury, United Kingdom. 2007.

[10] Brandrup, J., Bittner, M., Michaeli, W., and Menges, G. *Die Wiederverwertung von Kunststoffen.* Hanser, Munich. pp. 738–745, 1995.

[11] United States Environmental Protection Agency (EPA). *Advancing Sustainable Materials Management: Facts and Figures 2017. Assessing Trends in Material Generation, Recycling, Composting, Combustion with Energy Recovery and Landfilling in the United States.* Washington, D. C. 2019.

[12] Vault Consulting, LLC. *Resin Sales and Captive Use by Major Market.* Prepared for American Chemistry Council Plastics Industry Producers Statistics' Group. 2016.

[13] European Commission's Directorate-General Environment. *Plastic Waste: Ecological and Human Health Impacts.* Science for Environment Policy, In-depth Report. pp. 1–37, 2011.

[14] Franklin Associates. *Impact of Plastics Packaging on Life Cycle Energy Consumption & Greenhouse Gas Emissions in the United States and Canada.* Prepared for American Chemistry Council (ACC) and Canadian Plastics Industry Association (CPIA). 2014.

[15] Gerard, K. *Use of Plastic Materials in the Construction Industry.* Access Date: 2016/11/08. Available: *http://info.craftechind.com/blog/bid/380434/Use-of-Plastic-Materials-in-the-Construction-Industry*

[16] *nibusinessinfo.co.uk. Recycling construction materials – Recycling plastic from construction projects.* Access Date: 2016/11/08. Available: *https://www.nibusinessinfo.co.uk/content/recycling-plastic-construction-projects*

[17] Thibodeaux, J. *Recycling agricultural plastic is growing industry.* Access Date: 2016/11/08. Available: *http://www.greensourcedfw.org/articles/recycling-agricultural-plastic-growing-industry*

[18] Grossmann, E. *How can agriculture solve its $5.87 billion plastic problem?* Access Date: 2016/11/08. Available: *https://www.greenbiz.com/article/how-can-agriculture-solve-its-1-billion-plastic-problem*

[19] Barth, B. *3 Ways Farmers are Kicking the Plastic Habit.* Access Date: 2016/11/08. Available: *http://modernfarmer.com/2015/09/agriculture-plastic-waste/*

[20] Schwesig, A., Riise, B., and Ybbs, D. *PC/ABS recovered from shredded waste electrical and electronics equipment.* Paper presented at SPE Annual Technical Conference (ANTEC), Indianapolis, IN, 2016. In SPE ANTEC Conference Proceedings. pp. 1774–1779.

[21] Plastic Moulding. *Polyethylene.* Access Date: 2016/11/08. Available: *http://www.plasticmoulding.ca/polymers/polyethylene.htm*

[22] Polymer Properties Database. *A-Z Polymer Data.* Access Date: 2016/11/08. Available: *http://polymerdatabase.com/home.html*

[23] Holton, C. Dissolving the plastics problem. *Environmental Health Perspectives.* vol. 105, pp. 388–390, 1997.

[24] Illumin. *Recycling Plastics: New Recycling Technology and Biodegradable Polymer Development.* Access Date: 2016/11/08. Available: *http://illumin.usc.edu///recycling-plastics-new-recycling-technology-and-biodegradable-polymer-development/*

[25] Ruj, B., Pandey, V., Jash, P., and Srivastava, V. Sorting of plastic waste for effective recycling. *International Journal of Applied Science and Engineering Research.* vol. 4, pp. 564–571, 2015.

[26] Sanchez-Parra, K.M. *Effects of processing events on commodity thermoplastics.* Thesis, University of Wisconsin–Madison, Madison, WI. 2013.

[27] Basedow, A.A. M., Ebert, K.H., and Hunger, H. Effects of mechanical stress on the reactivity of polymers: shear degradation of polyacrylamide and dextran. *Die Makromolekulare Chemie.* vol. 180, pp. 411–427, 1979.

[28] Goldberg, V.M., and Zaikov, G.E. Kinetics of mechanical degradation in melts under model conditions and during processing of polymers—a review. *Polymer Degradation and Stability.* vol. 19, pp. 221–250, 1987.

[29] Peacock, A.J., and Calhoun, A. *Polymer Chemistry: Properties and Applications.* Hanser Gardner, Cincinnati, OH. 2006.

[30] Schmiederer, D. *Schonende Spritzgiessverarbeitung von Thermoplasten:* Univ. Erlangen-Nürnberg, Lehrstuhl für Kunststofftechnik. 2008.

[31] Zweifel, H. *Stabilization of Polymeric Materials.* Springer-Verlag, Berlin, Germany. 1998.

[32] Gugumus, F. Mechanisms of thermooxidative stabilization with HAS. *Polymer Degradation and Stability.* vol. 44, pp. 299–322, 1994.

[33] Moss, S., and Zweifel, H. Degradation and stabilization of high density polyethylene during multiple extrusions. *Polymer Degradation and Stability.* vol. 25, pp. 217–245, 1989.

[34] Frounchi, M. Studies on degradation of PET in mechanical recycling. *Macromolecular Symposia.* vol. 144, pp. 465–469, 1999.

[35] DeSousa, J. The effects of multiple heat histories on the mechanical properties of high-impact polystyrene. Paper presented at SPE Annual Technical (ANTEC), San Francisco, CA, May, 2002. In *SPE ANTEC Conference Proceedings.* vol. 3, p. 2925.

[36] Vilaplana, F., Ribes-Greus, A., and Karlsson, S. Degradation of recycled high-impact polystyrene. Simulation by reprocessing and thermo-oxidation. *Polymer Degradation and Stability.* vol. 91, pp. 2163–2170, 2006.

[37] Yarahmadi, N., Jakubowicz, I., and Enebro, J. Polylactic acid and its blends with petroleum-based resins: effects of reprocessing and recycling on properties. *Journal of Applied Polymer Science.* vol. 133, pp. 1–9, 2016.

[38] Case, R.M., Korzen, A.P., and Maclean, S.B. The effects of regrind loading levels and heat history on the properties of selected engineering polymers. Paper presented at SPE Annual Technical Conference (ANTEC), Dallas, TX, May 6–16. 2001.

[39] Eriksson, P.-A., Albertsson, A.-C., Boydell, P., Eriksson, K., and Månson, J.-A. E. Reprocessing of fiberglass reinforced polyamide 66: influence on short term properties. *Polymer Composites.* vol. 17, pp. 823–829, 1996.

[40] Eriksson, P. A., Albertsson, A. C., Boydell, P., Prautzsch, G., and Manson, J. A. E. Prediction of mechanical properties of recycled fiberglass reinforced polyamide 66. *Polymer Composites.* vol. 17, pp. 830–839, 1996.

[41] Gandhi, U., Goris, S., Osswald, T. A., and Song, S. *Discontinuous Fiber-Reinforced Composites: Fundamentals and Applications.* Hanser, Munich. 2020.

[42] I. V. K. AVK, *Handbuch Faserverbundkunststoffe - Grundlagen, Verarbeitung, Anwendungen.* Vieweg+Teubner Verlag, Wiesbaden, Germany. 2010.

[43] Vaidya, V. *Composites for Automotive, Truck and Mass Transit - Materials, Design, Manufacturing.* DEStech Publications, Inc., Pennsylvania, U. S. 2011.

[44] Albrecht, K. *Nachhaltige faserverstaerkte Kunststoffe im Spritzguss - Faserorientierung und Faserschaedigung im Experiment und Simulation,* Ph.D. thesis, University of Erlangen-Nuremberg, 2018.

[45] Thomason, J. L., and Vlug, M. A. Influence of fibre length and concentration on the properties of glass fibre-reinforced polypropylene: 1. Tensile and flexural modulus. *Compos. Part Appl. Sci. Manuf.* vol. 27, no. 6, pp. 477–484, 1996.

[46] Thomason, J. L. The influence of fibre length, diameter and concentration on the impact performance of long glass-fibre reinforced polyamide 6,6. *Compos. Part Appl. Sci. Manuf.* vol. 40, no. 2, pp. 114–124, 2009.

[47] Thomason, J. L., Vlug, M. A., Schipper, G., and Krikor, H. G. L. T. Influence of fibre length and concentration on the properties of glass fibre-reinforced polypropylene: Part 3. Strength and strain at failure. *Compos. Part Appl. Sci. Manuf.* vol. 27, no. 11, pp. 1075–1084, 1996.

[48] Thomason, J. L., and Vlug, M. A. Influence of fibre length and concentration on the properties of glass fibre-reinforced polypropylene: 4. Impact properties. *Compos. Part Appl. Sci. Manuf.* vol. 28, no. 3, pp. 277–288, 1997.

[49] Goris, S. *Characterization of the Process-Induced Fiber Configuration of Long Glass Fiber-Reinforced Thermoplastics,* Ph.D. thesis, University of Wisconsin-Madison.

[50] Lafranche, E., Krawczak, P., Ciolczyk, J.-P., and Maugey, J. Injection moulding of long glass fiber reinforced polyamide 66: Processing conditions/microstructure/flexural properties relationship. *Adv. Polym. Technol.* vol. 24, no. 2, pp. 114–131, 2005.

[51] Goris, S., Back, T., Yanev, A., Brands, D., Drummer, D., and Osswald, T. A. A novel fiber length measurement technique for discontinuous fiber-reinforced composites: A comparative study with existing methods. *Polym. Compos.* 39, pp. 4058–4070, 2018. doi:10.1002/pc.24466.

[52] Achilias, D. S., and Antonakou, E. V. Chemical and thermochemical recycling of polymers from waste electrical and electronic equipment, in *Recycling Materials Based on Environmentally Friendly Techniques,* IntechOpen, 2015.

[53] Pfaendner, R. Improving the quality of recycled materials - an overview of suitable additives, *Kunststoffe International,* 12/2015. pp. 41–44.

[54] Pospisil, J., Sitek, F. A., and Pfaendner, R. Upgrading of recycled plastics by restabilization - an overview. *Polymer Degradation and Stability,* 48, pp. 351–358, 1995.

[55] Osswald, T. A., and Menges, G. *Materials Science of Polymers for Engineers,* 3rd ed. Hanser, Munich. 2012.

4 Economic Analysis of Plastic Waste Handling

The first step to improve plastic waste handling is analyzing the procedures involved from an economic point of view. In order to do so, we will first review the *fundamentals of economic efficiency*. Afterwards, using these principles, we will analyze the economic efficiency of landfilling, incineration with energy recovery, and recycling, and then compare them.

4.1 Fundamentals of Economic Analysis

4.1.1 Economic Efficiency Calculation

In theory and practice, a variety of techniques to evaluate and compare economic efficiency exist. They are derived from classic capital budgeting, which considers the advantage of an investment with known acquisition costs and was designed for helping select the best long-term investments for financing. [1, 2]

Nowadays, capital budgeting also includes the valuation of companies and processes, via the so-called *economic efficiency calculation*. Its purpose is to measure the profitability of a company or process in order to derive an asking price for that company or process. It also enables the comparison of the efficiency of different processes with each other. [1] Therefore, the economic efficiency calculation can be used to compare landfilling, incineration with energy recovery, and recycling from an economic point of view.

In economic theory and entrepreneurial practice, there are two main approaches to making this calculation:

- *Static methods* determine the economic efficiency under constant framework conditions and explicitly consider only one time period, which is assumed to be representative for all such time periods. A typical time period under review is a year. The data for this time period is obtained by considering the whole planning period (in general the expected lifetime of an investment) and finally deriving the average cost and profit data for the time period under review. [1, 3]

- *Dynamic methods* also analyze the influence of the time of investment on the economic efficiency. By discounting payments and earnings using identical benchmarks via compound interest calculations, the impact of both the amount and the time of the investment on the profitability are considered. These methods are in general closer to reality, which is a clear advantage over static methods. If, for example, an investment in a new machine needs to be analyzed, dynamic methods enable one to include factors such as learning effects and decreasing manufacturing costs, which have a big impact on the economic efficiency. [1, 4, 5, 6]

The challenge of dynamic methods is the data collection and the uncertainty of future payments, earnings, and market behavior (e.g., inflation), which can lead to inaccuracies in the results. For this reason, static methods are still more frequently used. In practice, static and dynamic methods are often used in combination, since companies do not want to rely on only one of these methods alone. [1]

In the following, static methods will be used to analyze economic efficiencies of plastic waste handling.

4.1.2 Static Economic Efficiency Calculation

As mentioned in Section 4.1.1, static economic evaluations consider one average period (e.g., a year), which is assumed to be representative for the lifetime of the process. These methods focus on a pure financial measure of the profitability, which can be thought of in two ways:

- *Absolute profitability:* Investing in and/or running a business or process is economically better than not running it.
- *Relative profitability*: Investing in and/or running business A or process A is economically better than investing in and/or running business B or process B.

Static methods assume constant payments and earnings and do not take into account the changing value of cash over time. There are different static economic calculation methods, from which the profit comparison method is the most used one. Thus, this method will be further considered. [1, 3, 7]

4.1.3 Profit Comparison Method

The *profit comparison method (PCM)* computes the profit of plastic waste handling processes by subtracting the total costs of these processes from their total revenues. The two types of profitability considered are:

- *Absolute profitability:* the profit of a waste handling procedure
- *Relative profitability:* the profit of a particular waste handling procedure as compared to the profit of another waste handling procedure

The costs included in the calculation are: [3, 5]

- *Initial investment cost*
- *Capital cost*: depreciation and interest
- *Fixed costs*: lease rental charges and/or land costs, fixed salaries, and setup costs
- *Variable costs*: costs of material and tools; direct labor costs; energy, electricity, and water costs; maintenance costs; and taxes

For plastic waste handling, where the revenue comes from depends on the procedure: waste-to-energy (WTE) plants derive revenue from selling electricity produced, and recycling plants from selling the recycled pellets or plastic products.

After summing up all costs and revenues for each plastic waste handling procedure, the average cash flows for one representative time period or unit need to be determined to estimate the profit of each procedure. Here the absolute and relative profitability will be determined for handling of **1 ton (t) of plastic waste**. [3]

4.2 Economic Analysis of Landfilling

As described in Section 1.2.1 landfills are required to comply with federal regulations, which include among other things location restrictions, design specifications, operating standards, and closure requirements. These regulations have a great influence on the cost to build, operate, and finally close landfills. [8, 9]

Besides regulatory issues, *landfill costs* are site specific; thus, they vary based on factors such as terrain, soil type, climate, site restrictions, and amount of waste disposed. For economic analyses, the costs of landfills are mainly grouped into four categories: [9, 10, 11, 12, 13]

- Construction costs
- Operations costs
- Closure costs
- Postclosure and maintenance costs

To analyze the profit of landfilling, a large landfill with a footprint of 365.76 m × 365.76 m (1,200 ft × 1,200 ft, approximately 33 acres) will be analyzed. As already mentioned in Section 1.2.1, the trend in the United States is towards large landfills. The average landfill has a yearly disposal of 70,000 t (for a total of around 140 Mt per year on 1,269 landfills). The landfill considered here has a yearly disposal of 200,000 t, or 500 to 600 t daily.

Although most of the literature values were based on the U.S. customary or imperial system (feet [ft], acres), the Appendix (Chapter 8) includes calculations both for the imperial and metric system. Table 4.1 summarizes important assumptions that are required for the economic analysis. [11, 12, 13]

Table 4.1 Assumptions for Economic Analysis of Landfill

Lifetime [years]	11	
Postclosure care period [years]	30	
Final surface grades [acres] | [hectares (ha)]	34.00	13.76
Bottom of landfill [acres] | [hectares (ha)]	33.50	13.56
Disposal capacity constructed acre [yd^3/acre] | [m^2/m^3]	60,000	19.39
Total capacity [yd^3] | [m^3]	4,000,000	3,058,104
In-place density of waste [t/yd^3] | [t/m^3]	0.55	0.72
Weight capacity per acre [t/acre] | [t/m^2]	33,000	49.42
Waste per day [t]	500–600	
Waste per year [t]	200,000	
Total weight capacity [t]	2,227,500	
Total overall weight of landfill (11 years) [t]	2,200,000	

To calculate the profit of landfilling, the costs must be considered first. Therefore, the total overall costs of the 13.35 ha (33 acres) landfill will be summarized and finally divided by total weight of municipal solid waste (MSW) disposed over the landfill's lifetime (2,200,000 t).

The first cost category is *construction*, which is divided into that for the support facility and that for the liner construction. Based on the 365.76 m × 365.76 m (1,200 ft × 1,200 ft) footprint, the landfill's access road has a length of approximately 381.00 m (1,250 ft), in total 1,524.00 m (5,000 ft). Including possible setbacks for the landfill from the property line required by the state regulatory agency and additional area required for other facilities, the security fence has a length of 1,828.82 m (6,000 ft). The costs per foot or per meter of road and fence are given in the Appendix in Table 8.1 (imperial) and Table 8.2 (metric), respectively. The space and costs for support buildings, including office buildings, maintenance buildings, shacks, and tool sheds, are found in Table 8.1 (imperial) and Table 8.2 (metric). In addition, every landfill plant needs one modular truck scale and an associated computer system. [13]

The bottom of the landfill with an area of 13.56 ha (33.5 acres) needs a liner and leachate system. After performing a site survey, the bottom area first needs to be cleared and grubbed. Once completed, liner construction grades and elevations are established by excavation. It is assumed that the landfill has a minimal structural fill berm constructed along the landfill's perimeter to provide anchoring for the liner elements and structural toe stability for the final waste slope. After establishing the base grades, the landfill's liner and leachate management system, including clay liner, geomembrane, geocomposite, and granulator soil, is constructed. Besides these physical acts of construction and installation, management and quality oversight are required, typically done by independent third-party consultants. The detailed statement of costs is also found in Table 8.1 (imperial) and Table 8.2 (metric).

On average, the overall *construction costs* for the completed landfill total **$21,582,950**, as seen in Table 4.2. [11, 13]

Table 4.2 Construction Costs of Landfill

	Minimum	Maximum	Average
Support facility construction [$]	1,175,300	1,790,600	1,482,950
Liner construction [$]	11,256,000	28,944,000	20,100,000
Total construction costs [$]	**12,431,300**	**30,734,600**	**21,582,950**

Operating costs include staffing, utilities, equipment operations, leachate disposal and treatment, scale operations, paperwork, record keeping, billing, engineering staffing, environmental monitoring, and daily cover applications. Assuming waste processing of 200,000 t per year, which is equivalent to waste receipt of 500 to 600 t per day, this landfill would at minimum require a front-end loader for onsite hauling of bulk material and small construction tasks (e.g., CAT 950H), a bulldozer equipped with a trash rack to spread dirt and waste (e.g., CAT D7), and a steel-wheeled compactor to compact the waste and achieve maximum possible in-place density (e.g., CAT 826G). Additional equipment such as water spray trucks (for holding dust down), a scraper, a backhoe, several pickup trucks, and a road grader would also be needed. [14, 15, 16]

The salaries of all employees are based on average U.S. values. Summarizing all operations costs, the yearly operations costs amount to **$1,578,300**, or **$17,361,300** after 11 years, as presented in Table 4.3. Detailed operations cost calculations are shown in Table 8.3 in the Appendix. [11, 13, 17, 18, 19, 20, 21]

Table 4.3 Operations Costs of Landfill

Operations [$]	1,368,300
Leachate collection and treatment [$]	30,000
Environmental sampling and monitoring [$]	60,000
Engineering services [$]	120,000
Total operations costs per year [$]	**1,578,300**
Total operations costs of landfill (11 years) [$]	**17,361,300**

After use for disposing of waste for 11 years, the landfill must be closed. Due to sloping, the final surface grades needing capping and covering are approximately 13.76 ha (34 acres). The closure process starts with surveying the surface to receive a final cap and cover. Once this step is done, construction of the final cap and cover begin, which includes a geomembrane cap and vegetative soil, seed, mulch, and fertilizer. To prevent air and soil pollution, run-off and gas control systems are installed. Besides these physical acts of construction and installation, management and quality oversight are required. A detailed cost breakdown is shown in Table 8.4 (imperial) and Table 8.5 (metric) in the Appendix. Summarizing the mentioned categories, *closure costs* of the considered landfill range from **$7,718,000** to **$11,084,000**, with an average total cost of **$9,401,000**. [11, 12]

Once a landfill is closed, RCRA prescribes the care for and maintenance of landfills for a minimum of 30 years. *Postclosure and maintenance costs* are mainly divided into site security maintenance, landfill cover, and mechanical systems maintenance, monitoring wells and gas probes, and environmental monitoring. For 1 year, these costs range between $72,182 and $100,232 for the complete landfill, which is shown in Table 8.6 (imperial) and Table 8.7 (metric) in the Appendix. Over 30 years, the complete postclosure and maintenance costs add up to **$2,586,210**.

To finally calculate the cost per ton of plastic waste, the total costs of all categories are summed up and divided by the total waste disposed of in the landfill (2,200,000 t).

Table 4.4 Total Costs of Landfill

Category	Minimum	Maximum	Average
Construction costs [$]	12,431,300	30,734,600	21,582,950
Operations costs [$]	17,361,300	17,361,300	17,361,300
Closure costs [$]	7,718,000	11,084,000	9,401,000
Postclosure/maintenance costs [$]	2,165,460	3,006,960	2,586,210
Total costs [$]	**39,676,060**	**62,186,860**	**50,931,460**
Total costs per ton [$/t]	**18.03**	**28.27**	**23.15**

As shown in Table 4.4, the total average cost of landfilling 1 t of plastic waste is **$23.15**.

The revenue from landfilling 1 t of plastic waste is **$0.00**. Nevertheless, it is possible for landfill owners to receive revenue by selling carbon credits, selling electricity generated from landfill gas (LFG) to the local power grid, or selling LFG to a direct end user or pipeline. However, the revenue received represents a negligible percentage of the operating costs of a landfill and is therefore set to $0.00. [9, 22]

Applying the profit comparison method, the last step is to calculate the overall profit of landfilling 1 t of plastic waste.

Table 4.5 Total Profit per Ton of Plastic Waste for Landfill

	Minimum	Maximum	Average
Revenues per ton [$/t]	0.00		
Costs per ton [$/t]	18.03	28.27	23.15
Profit per ton [$/t]	**– 18.03**	**– 28.27**	**– 23.15**

Subtracting the cost per ton from the revenue per ton, the average profit of landfilling plastic waste is **–$23.15 per ton**. This means landfilling is not absolutely profitable (profit < $0.00).

The figure of –$23.15 per ton is the costs for a big landfill. The bigger the landfill, the smaller the relative costs. If the average waste per day were for example only 450 t, but the landfill is constructed for 550 t per day, the yearly waste would be around 165,000 t and after 11 years only 1,815,000 t. The landfill would have to be closed due to national regulations; the total costs would be the same and the costs per ton would be on average **$28.27**.

Another big cost factor that is not included in the calculations is the costs for cleaning up landfills in cases of accidents. Since most of the landfills are not cleaned up afterwards due to high costs, the experiences with these costs are comparably low. But some examples can be found and allow us to éstimate this cost. In 2008, a dike failed at the Tennessee Valley Authority's Kingston Fossil Plant. A total of 5.4 million yd^3 of coal ash cascaded into the Emory and Clinch Rivers and smothered about 300 acres of land. Up until 2015, about **$1.2 billion** were spent to clean up the area, and more will probably follow. This is about 19 times more than the maximum costs for the landfill we are considering here. [23]

In the future, costs for landfilling will probably increase substantially. Since land is becoming a more and more scarce resource, the costs for land and construction are likely to rise. This trend can be seen in the average *tipping fee,* also called the gate fee, of landfills in the United States. Between 1985 and 2018, the average tipping fee increased from **$8.22** to **$55.00** per ton of waste. The increase was likely caused by states implementing RCRA Subtitle D regulations or their state's equivalent. Furthermore, China's import ban for plastic waste, which will be further analyzed in Chapter 7, leads to an increased average tipping fee. [13, 22, 24, 25, 60]

4.3 Economic Analysis of Incineration with Energy Recovery (Waste-to-Energy Facilities)

As already mentioned in Section 1.2.2, the most widespread type of waste-to-energy plants in the United States are mass-burn facilities (64 out of 84), and therefore we use an example of one for this profitability analysis. A plant diagram of a typical mass-burn facility is shown in Figure 1.3. [26]

The profit analysis of mass-burn facilities is mainly split into three categories: [27]

- Investment costs
- Operating and maintenance costs
- Revenues from sale of heat and electricity generated

The actual profit for a waste incineration plant depends on a wide range of factors, especially the capacity of the plant and the type of energy production. The facility considered here is a co-generation facility and produces both heat and electricity. Since new WTE facilities have an expected lifetime of 35 years, this period was assumed for the following analysis. As for the landfill analysis (Section 4.2), the complete profit of plastic combustion over the facility's lifetime is summed up and divided by the total plastic waste burned (3,500,000 t). The important general assumptions of this analysis are shown in Table 4.6, and in more detail in Table 8.8 in the Appendix. [28, 29]

Table 4.6 Assumptions for Economic Analysis of Waste-to-Energy (Mass-Burn) Plant

Lifetime [years]	35
Yearly waste capacity [t]	100,000
Total waste capacity (35 years) [t]	3,500,000
Steam generator efficiency [%]	80
Percentage of plastic in MSW [%]	12.8
Dollar-Euro exchange rate [$/€]	1.2125
Purchasing power parity (PPP) conversion factor of Croatia	0.6

Data for *investment costs* was taken from a waste-to-energy plant in Croatia, since this was the only comprehensible data available. To realistically transfer this data to the U.S. market, it first needed to be converted from Dollars to Euros at an average for the last 20 years of **$1.21** per Euro. [30]

Furthermore, the prices were adjusted using the *purchasing power parity (PPP)* conversion factor, which represents the number of units of a country's currency required to buy the same amount of goods and services in that country as a U.S. Dollar would buy in the United States. The PPP conversion factor of Croatia was consistently 0.6 over the last 15 years and therefore is used for these calculations. [31]

Investment costs vary substantially with respect to several factors, including the design of the WTE plant, its capacity, and the local infrastructure. Construction of a road, a weighing area, and waste reception storage is necessary. The combustion chamber-steam generator system includes a system for feeding waste into the combustion chamber, an air supply for the combustion chamber, a combustion chamber, ash removal and storage, flue gas channels, and a steam generator with steam output. The water and steam system consists of a water treatment facility, an air-cooled condenser, and a condensation turbine. Flue gas cleaning, which represents an important part of the overall waste combustion process, is ensured by a flue gas cleaning system involving a semi-dry treatment, a bag filter, and a selective noncatalytic reduction (SNCR) system. Additional initial costs are design, construction, electromechanical installations, and other smaller costs. Summarizing all these costs, the total investment for the WTE plant amounts to **$113,781,924**, as shown in Table 4.7 and in more detail in Table 8.9 in the Appendix. [29]

Table 4.7 Investment Costs of Waste-to-Energy Plant

Infrastructure and waste storage [$]	9,296,569
Combustion system and steam generator [$]	39,409,370
Water and steam system [$]	16,167,946
Design [$]	4,041,986
Construction [$]	14,146,953
Electromechanical installations [$]	10,104,966
Other investment costs [$]	12,125,960
Flue gas cleaning system [$]	8,488,172
Total investment costs [$]	**113,781,924**

Operating and maintenance costs are analyzed annually and divided into personal salaries, operation and maintenance costs as well as heat and electricity costs. The plant runs three shifts daily. Per shift, 15 workers, 5 environmental engineers, 3 maintenance engineers, and 1 facility manager are required. Salaries used for the calculations are average U.S. salaries. The annual salaries of all personnel amount to $3,570,000. Maintenance costs of the combustion system are proportional to the waste flow and estimated at 3% of total investment costs. Additional annual machine and emission costs are process water, natural gas, bottom ash disposal, and bag filter residues costs. Total annual machine and emission costs are $10,190,980. Assuming electricity consumption of 0.1 MW/t_{waste} with a price of $0.1027 per kWh and a heat consumption of 0.05 MW/t_{waste} with a price of $0.0011 per kWh, the annual heat and electricity costs are $105,454. Operating and maintenance costs add up to $13,866,434 per year and $485,325,214 after

35 years, as presented in Table 4.8. A more detailed calculation is provided in Table 8.10 in the Appendix. [21, 29, 32, 33, 34, 35]

Table 4.8 Operating and Maintenance Costs of Waste-to-Energy Plant

Personnel salaries per year [$]	3,570,000
Machine and emission costs per year [$]	10,190,980
Heat and electricity costs per year [$]	105,454
Yearly operating and maintenance costs [$]	**13,866,434**
Overall operating and maintenance costs [$]	**485,325,214**

The revenues from plastic combustion come from selling the electricity and heat produced by these plants. These revenues depend on the average lower heating factor of burned material. Compared to general MSW, the average lower heating value of plastics is high. Based on the composition of plastics in the MSW stream (Table 2.1), the lower heating value of plastics in waste is **36.16 MJ/kg** (detailed calculations are found in Table 8.11 in the Appendix). The general MSW has an average lower heating value of only **14.98 MJ/kg** (detailed calculations are found in Table 8.12 in the Appendix). All values and assumptions necessary for the revenue calculations are presented in Table 4.9. [36, 37]

Table 4.9 Assumptions for Revenue Calculations from Sale of Energy Generated in Waste-to-Energy Plant

Percentage heat of produced energy [%]	75
Percentage electricity of produced energy [%]	25
Energy per cubic meter in gas [kWh/m^3]	12.5
Average lower heating value MSW [MJ/kg]	14.98
Average lower heating value plastic [MJ/kg]	36.16
Costs/Revenue per kWh electricity [$/kWh]	0.1027
Cost/Revenue per cubic ft gas [$/ft^3]	0.0039
Cost/Revenue per cubic meter gas [$/m^3]	0.1377

Assuming a percentage of plastics in MSW of 13.2%, 13,200 t of plastic waste (out of 100,000 t) are incinerated annually. Theoretically, the energy produced through plastics combustion is therefore 462,833,408 MJ per year, 75% of which produces heat and 25% of which produces electricity. Assuming a steam generator efficiency of 80%, the electricity produced yearly amounts to 32,141,209 kWh, and the heat produced yearly to 96,423,627 kWh. [38]

The average price of electricity for customers in the United States in December 2019 was $0.1027 per kWh. If WTE plants produce 32,141,209 kWh/year, annual revenues from selling electricity generated by plastics combustion are $2,561,011. Supposing an energy content of 12.5 kWh per cubic meter gas, the generated heat would produce 6,171,112 m^3 of gas per year. With an average gas price of $0.1377 per m^3, annual revenues of selling gas produced by plastics combustion are $849,931. [35]

Summing up the revenues of the sale of both the electricity and heat and multiplying it by the plant lifetime of 35 years, total overall revenues of incinerating plastics add up to **$119,383,002**, as shown in Table 4.10.

Table 4.10 Total Revenue from Sale of Energy Generated in Waste-to-Energy Plant

Yearly revenue for electricity [$]	2,561,011
Yearly revenue for heat [$]	849,931
Yearly revenue of plastic waste combustion [$]	**3,410,942**
Total revenue of plastic waste combustion [$]	**119,383,002**

Finally, to calculate the profit from combustion of 1 t of plastic waste, the costs need to be subtracted from the revenues. Therefore, the investment, operations, and maintenance costs are added and divided by total waste burned over the plant's lifetime (3,500,000 t). Then the total revenues of plastic waste combustion are divided by the total amount of plastic burned over 35 years (448,000 t). These calculations result in costs of $171.17 and revenues of $266.48 per ton of plastic waste, which leads to a profit of **$95.31** per ton of plastic waste burned, as shown in Table 4.11.

Table 4.11 Profit per Ton from Combustion of Plastic Waste

Revenues per ton of plastic waste burned [$/t]	266.48
Costs per ton of plastic waste burned [$/t]	171.71
Profit per ton of plastic waste burned [$/t]	**95.31**

The profit of **$95.31** per ton of plastic waste burned is an absolute profit (profit > $0.00). As already mentioned, this value can only be reached theoretically. The calculation assumed that no plastic waste was recycled or treated any other way. Furthermore, in a real scenario, plastic waste would not be combusted separately from other waste. But due to its high average lower heating value, plastic waste combustion results in high energy production and is absolutely profitable.

To review and check the WTE-plant analysis, the profit of combustion of 1 t of MSW was calculated and compared to the average tipping fee of combustion of 1 t of waste in the United States. The costs of burning 1 t of MSW are equivalent to the costs of incinerating 1 t of plastic waste ($171.71). To determine revenues of MSW combustion, an average lower heating value of 14.98 MJ/kg was calculated (see Table 8.12 in the Appendix). This resulted in a yearly revenue (from heat and gas) of $10,808,957, or **$378,313,500** after 35 years. Dividing the total revenues by the total incinerated MSW over the plant's lifetime, the revenue from combustion of 1 ton of MSW is $110.38.

Subtracting costs from revenues, the total profit of MSW combustion is –$60.80 per ton of MSW.

Table 4.12 Profit per Ton from Combustion of Municipal Solid Waste (MSW)

Revenues per ton of MSW [$/t]	110.38
Costs per ton of MSW [$/t]	171.71
Profit per ton of MSW [$/t]	**–60.80**

The average tipping fee for combustion of 1 t of MSW in the United States is **$64.96 per ton** (Table 8.13) and represents approximately the costs of waste combustion. This fee only differs by **$4.16** (6.4%) from the calculated value and therefore indicates reasonable assumptions and calculations of the previous WTE plant analysis. [39, 40, 41]

A factor not being included in the profit analysis is the case of rebuilding a plant: a new plant could be built on the site of the existing WTE plant, which reduces the capital cost for land in the new facility to zero. This is a big economical advantage, especially compared to landfill plants, the land from which cannot be used again after closure for a long postclosure period. [42]

4.4 Economic Analysis of Plastics Recycling

The third plastic handling method is *recycling*. Figure 4.1 shows the plastics recycling process beginning and ending with the consumer.

After plastic is thrown away and collected within the single recycling stream, material recovery facilities (MRF) sort the material into the seven plastic recycling groups shown in Table 3.1. [43]

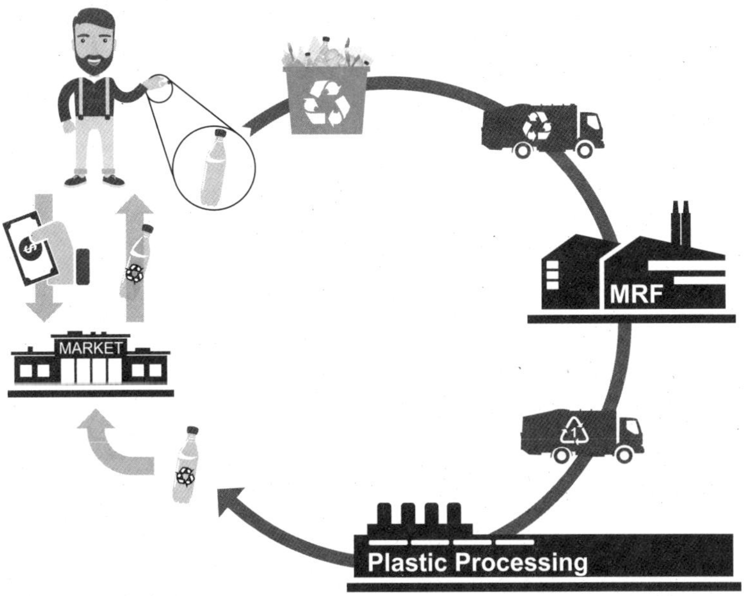

Figure 4.1 Overview of the plastics recycling process

The higher the resin number, the harder the recycling of that plastic. In the MRF, plastics are sorted into three groups: first, number 2 plastic (HDPE) is sorted out manually and next, number 1 plastic (polyethylene terephthalate [PET]) is optically separated from numbers 3 to 7 plastics with an automated IR detector. Due to its good recycling properties, number 1 plastic is the most valuable and at present the only plastic in the United States that is recycled, reprocessed, and resold as pure material in significant quantities. Consequently, the economic analysis of the plastic recycling process in this book will be based on PET recycling.

There are three steps to the economic analysis of the profitability of PET recycling that will be analyzed in the following:

- The cost of material recovery in the MRF (Section 4.4.1)
- The cost of plastic reprocessing (Section 4.4.2)
- The revenue derived from selling recycled PET (Section 4.4.3)

4.4.1 Materials Recovery Facility Costs

Plastic waste is collected by dump trucks together with all the other recyclable materials such as metal, paper, and glass waste and brought to a *materials recovery facility (MRF)*. A MRF is a processing facility where waste is sorted and prepared for additional processing. Configurations of the MRF processing line vary depending upon how waste is collected and received at MRFs. The four different ways of collecting waste are source-separated, dual-stream, single-stream, and mixed-waste collection. Due to advances in automated processing equipment, less upfront separation is required; thus, many source-separated MRFs gave way to dual-stream MRFs, which in turn are being replaced by single-stream MRFs in some locations. For this reason, a semi-automated MRF for single-stream recycling is considered for the analysis of plastics recycling in this book. A flowchart of the process is shown in Figure 4.2. The solid lines represent the flow of PET. [44]

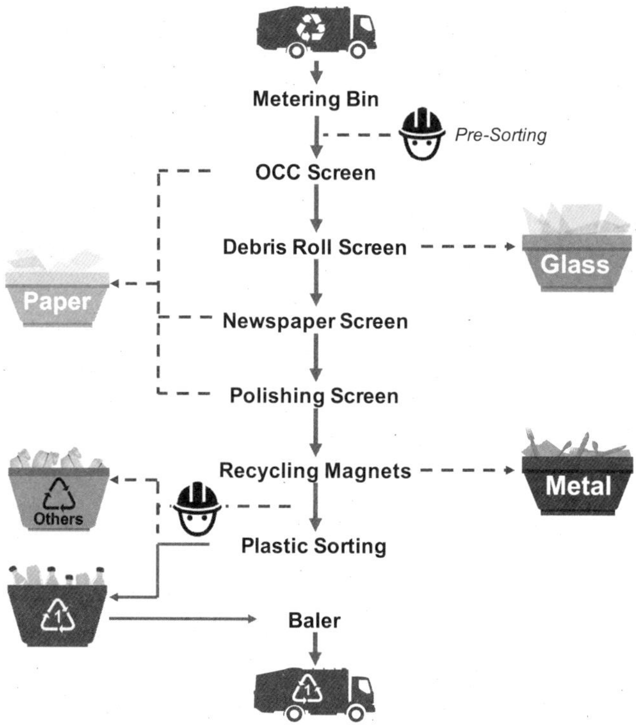

Figure 4.2 Schematic of materials recovery facility (MRF). OCC screen, old corrugated cardboard screen

As trucks arrive at the MRF, they unload recyclable waste onto the tipping floor, where the material is stored until processing. Usually, the tipping floor is sized to hold at least two days of incoming waste to allow a buffer against unscheduled equipment downtime and to provide sufficient material for the MRF to operate during an extra shift. The dumped waste is grabbed by a loader to fill a *metering bin;* this serves to produce a constant flow. This constant feed rate prevents surges, allows for more efficient manual sorting in the presort area, and maximizes the efficiency of automated equipment encountered later in the processing line. During the manual presorting, large contaminants, bulky recyclables, and items that could damage downstream sorting equipment or pose a threat to personnel are removed. [44, 45]

As material leaves the presort station, it is transferred to the next machine, the so-called *old corrugated cardboard (OCC) screen*. It is composed of several rotating star-shaped disks, which separate the OCC from all the other materials by using size and dimension. All non-OCC materials fall through the disks onto the next screen. Below the OCC screen is a *debris roll screen*, which acts as a glass breaker. It breaks glass into quarter-sized pieces, which then fall through the debris roll screen and go into the glass trailer for transportation to a glass recycler.

After OCC and glass are removed, residual waste is routed to a *newspaper screen*, which uses the same concept as the OCC screen. The difference between these screens is the distance between the disks. Disks of the newspaper screen are closer together, so that the newspaper material stays on the disks and the smaller materials fall through. The *polishing screen* has even closer disks, which separate remaining paper waste from other materials. [44, 45]

OCC, newspapers, and other paper materials are finally routed to a paper platform, where the paper waste is checked manually and contaminants, which "acted" like paper when passing through the three paper screens, are removed from the paper waste before paper is sent out to a paper recycler. [44, 45]

A second system, called an *eddy current separator*, induces an electrical current in conducting metals such as aluminum. This in turn induces a magnetic field in these materials, repelling the aluminum and thus removing it from the waste stream. [44, 45]

The only materials remaining in the system are numbers 1 to 7 plastics (Table 3.1). In a first step, number 2 plastics (HDPE) are taken out manually (see Section 3.3.1 Manual Sorting). Afterwards, the residual plastics are sorted by an *optical sorter* (see Section 3.3.2.3 Near Infrared Sorting). This machine shoots a ray of light into the plastic and analyzes how the light reflects back to determine if the item is a number 1 (PET) or numbers 3 to 7 plastics. The optical system shoots a controlled blast of air on the item, which directs it to either the number 1 or the numbers 3 to 7 plastic storage area. Nonplastic materials do not receive any air shot and fall down to the residue conveyor, ending in the trash. [44, 45]

Sorted PET is then in the interim stored in a separate material bunker until the loose material is compacted by a *baler* into a large, bricklike bale to prepare the material for shipping to a plastic processing company. [44, 45]

The costs of the MRF process are split into two categories, investments and operating costs. As in the landfill and WTE facility analysis, all costs over the plant's lifetime are summarized first and afterwards converted to a cost per 1 t of plastic waste processed. The assumptions for the economic analysis are presented in Table 4.13 and in greater detail in Table 8.14 and Table 8.15 in the Appendix. [46, 47]

Table 4.13 Assumptions for Economic Analysis of Materials Recovery Facility

Lifetime [years]	10
Yearly working hours [h]	4,160
Waste handled per hour [t]	30
Yearly waste handling [t]	120,000
Total waste handling (10 years) [t]	1,200,000
Residues rate [%]	2
Correctness of manual sorting [%]	91

Since the factory runs 52 weeks per year, 5 days per week in 2 shifts with an effective working time of 8 hours each shift, the yearly working hours add up to 4,160 hours. Since 30 tons are handled per hour (restricted by the capacity of the metering bin), around 120,000 tons of waste is handled per year, resulting in a total waste handling of 1,200,000 tons of waste after 10 years. [47]

Investment costs for the MRF are divided into buildings and site, machine, and equipment costs. For an 80,000 ft^2 large area, initial costs for the land and site work are $1,395,000, and buildings costs for the scale house and MRF building are $9,200,000. Construction, planning, and surveying costs amount to $3,500,000, so the total investment costs for the buildings and site add up to $14,095,000. The investment costs of all machines for the MRF (Figure 4.2) amount to $2,163,000. Additional equipment are conveyors, rolling stock (e.g., front-end loaders and forklifts), and collection cars, the costs of which add up to $1,400,000. Summing up the investment costs from these three categories, the total investment costs of an MRF are **$17,658,000**, as shown in Table 4.14 and in more detail in Table 8.16 in the Appendix. [46, 47, 48]

Table 4.14 Total Investment Costs of Materials Recovery Facility

Building and site investment costs [$]	14,095,000
Machine investments costs [$]	2,163,000
Additional equipment investment costs [$]	1,400,000
Total investment costs [$]	**17,658,000**

Operating and maintenance costs of the MRF are divided into seven different categories. The yearly operating and maintenance costs are $8,451,851, so overall operating and maintenance costs of $84,518,517 after 10 years, as shown in Table 4.15 and in more detail in Table 8.17 in the Appendix.

Table 4.15 Total Operating and Maintenance (O&M) Costs of Materials Recovery Facility

Personnel salaries per year [$]	2,552,000
Facility costs per year [$]	343,000
Machine O&M costs per year [$]	130,602
Conveyor O&M costs per year [$]	12,392
Rolling stock O&M costs per year [$]	897,056
Residues costs [$]	96,000
Transportation and collection costs [$]	4,420,800
Yearly O&M costs [$]	**8,451,851**
Overall O&M costs (10 years) [$]	**84,518,517**

The largest group of personnel is the sorters for presorting, manual HDPE sorting, and quality control. A detailed breakdown of personnel is provided in Table 8.17 in the Appendix. The yearly salaries amount to $2,552,000. The salaries used for the calculation are average yearly U.S. salaries. [18, 19, 20, 21, 48]

Facility costs include consumables, insurance, administration, and baling wire and amount yearly to $343,000. Operating and maintenance costs of machines as well as conveyors involve yearly maintenance and electricity consumption ($0.1027 per kWh) and add up to yearly $142,995. The rolling stock has a high consumption of diesel, so the yearly costs are $897,056. Assuming that 2% of the waste is residues and that all residues are landfilled, yearly costs for residues are $96,000. The largest part of operating and maintenance costs is generated by transportation and collection: assuming transportation and collection costs of $36.84 per ton, yearly costs for 120,000 t are $4,420,800.

Summarizing the investment costs and overall operating and maintenance costs, the total MRF costs are **$102,176,517**, or **$85.15** per ton of plastic waste (Table 4.16).

Table 4.16 Total Costs of Materials Recovery Facility

Investments costs [$]	17,658,000
Operating and maintenance costs [$]	84,518,517
Total costs (10 years–1,200,000 t) [$]	**102,176,517**
Costs per ton of plastic waste [$/t]	**85.15**

4.4.2 Plastic Reprocessing Costs

After PET is baled in the MRF, the bales are transported to a *plastic reprocessing facility*, where they are further treated, as schematically presented in Figure 4.3.

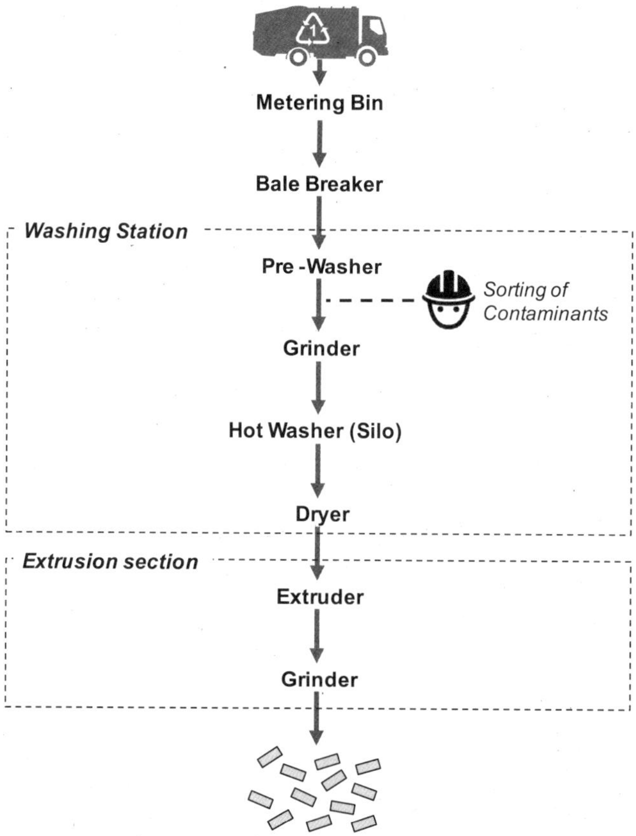

Figure 4.3 Schematic of plastic reprocessing facility

From the tipping floor, PET bales are grabbed by a loader and laid into a metering bin, which constantly meters the plastic waste into a bale breaker. The bale breaker dismembers the PET bales into individual free flowing items (e. g., food containers and bottles). [49, 50]

The individual items are conveyed to a washing station. After a short prewashing to remove labels and dirt from the outside of the items and a manual hand sorting of contaminants, PET items are ground into flakes by a wet granulator. These ground flakes are transported to a silo for hot washing, which removes the last dirt and glue. In a final step at the washing station, these clean flakes are dried. [49, 50]

The dry flakes are metered to an extrusion section, which melts, vacuum degasses, and finally grinds the clean washed flakes to produce food-grade natural PET pellets. These are stored in big sacks and are ready to be used to make new products. [49, 50]

As for the MRF process, costs of the PET process are split into two categories, investments and operating costs, summarized over the factory's lifetime and then divided by the tons processed over the factory's lifetime to yield the cost per ton of plastic waste. The assumptions for the economic analysis are presented in Table 4.17 as well as Table 8.18 in the Appendix.

Table 4.17 Assumptions for Economic Analysis of Plastic Reprocessing Facility

Lifetime [years]	10
Yearly working hours [h]	6,240
Yearly PET capacity [t]	15,000
Total waste capacity (10 years) [t]	150,000
Separation efficiency [%]	91
Dollar to British Pound conversion rate [$/£]	1.46

Investment costs include building and site, machine, and additional equipment costs. Since the source of some of the investment costs were the United Kingdom, a Dollar to Pound conversion rate of 1.46 $/£ (as of April 26, 2016) was used for the calculations. Including costs for design and project management, civil engineering, land, and site work, the building and site investments added up to $4,140,000. The metering bin, bale breaker, washing station, and PET extrusion section costs were $25,852,220. These costs included conveyors and installation costs. For auxiliary equipment, further costs of $150,000 were incurred. Summed up, the investment costs for the PET reprocessing plant were **$30,142,220**, as shown in Table 4.18 (and in more detail in Table 8.19 in the Appendix). [47, 49, 51]

Table 4.18 Total Investment Costs of Plastic Reprocessing Facility

Building and site investment costs [$]	4,140,000
Machine investment costs [$]	25,852,220
Additional equipment investment costs [$]	150,000
Total investment costs [$]	**30,142,220**

Operating and maintenance costs include personnel salaries, maintenance, machine operating, and rolling stock operating costs. Assuming the factory runs 52 weeks per year, 5 days per week in 3 shifts with an effective working time of 8 hours each shift, the yearly working hours are 6,240 hours and result in personnel salaries of $2,193,290 per year, which are mainly for sorters. Maintenance costs of the machines are assumed as 5% of the investment costs and therefore amount to $1,292,611. Operating costs are caused by energy and water consumption especially for washing and extruding, which consume a high amount of energy, and lead to operating costs of $1,054,603 per year. Costs for diesel for the rolling stock are $111,507 per year. These four categories result in total operating and maintenance costs of **$4,652,011** per year, so **$46,520,110** after 10 years, as presented in Table 4.19. Detailed calculations of operating and maintenance costs are provided in Table 8.20 in the Appendix. [19, 20, 21, 47, 49]

Table 4.19 Total Operating and Maintenance Costs of Plastic Reprocessing Facility

Personnel salaries per year [$]	2,193,290
Maintenance costs per year [$]	1,292,611
Machine operating costs per year [$]	1,054,603
Rolling stock operating costs per year [$]	111,507
Yearly O&M costs [$]	**4,652,011**
Overall O&M costs (10 years) [$]	**46,520,110**

Summarizing investment and overall operating and maintenance expenses, processing 150,000 tons of PET in 10 years costs **$76,662,330**, or **$511.08** per ton of PET, as it is presented in Table 4.20. Since only 14.16% of plastic waste is PET (Table 2.1), the costs of plastic processing of 1 t of plastic waste are **$72.37**.

Table 4.20 Total Costs of Plastic Reprocessing Facility

Investments costs [$]	30,142,220
Operating and maintenance costs [$]	46,520,110
Total costs (10 years; 150,000 t) [$]	**76,662,330**
Costs per ton of PET [$/t]	**511.08**
Costs per ton of plastic waste (14.16% PET) [$/t]	**72.37**

4.4.3 Revenues from Selling Recycled Plastic

The revenues from the PET recycling process come from selling the PET pellets produced to companies that produce new PET products for different applications, such as carbonated drink bottles (blow molding), food containers (thermoforming), or fibers (spinning).

In March 2020, the market price of 1 kg of PET pellets in the United States was on average $1.28. [52]

Calculating the revenues of plastic recycling, it needs to be remarked that only 14.16% of the plastic waste is PET (Table 2.1), so only this 14.16% of 1 t of plastic waste is recycled in a best-case scenario, which means 141.6 kg out of 1 t. Furthermore, the efficiencies of both optical sorting during material recovery (89.18%) and manual sorting during PET processing need to be considered. [47, 49, 53]

With an average price of $1.01/kg for recycled PET, revenues from recycling 1 t of plastic waste are $146.94, as shown in Table 4.21.

Table 4.21 Revenues from Recycling per Ton of PET Plastic Waste

Amount of PET in 1 t of plastic waste [kg]	141.60
Separation efficiency of plastic recycling [%]	89.18
Separation efficiency of PET processing [%]	91.00
Price of recycled PET [$/kg]	1.28
Total revenues per ton of plastic waste [$]	**146.94**

As also shown in Table 2.1, small amounts of the other plastics are recycled. Currently, these other plastics produce close to no revenue due to a lack of reprocessing procedures for them in the United States. They are normally landfilled and so their revenues are not considered in this analysis. [53]

4.4.4 Profitability

Finally, to calculate the profitability, the total costs per ton of waste need to be subtracted from its revenues.

The costs of both MRF and PET processing add up to $157.52 per ton of plastic waste, and the revenues of recycling 1 t of plastic waste are $146.94. Thus, the absolute profit from recycling 1 t of plastic waste is **−$10.58** (see Table 4.22).

Table 4.22 Profit from Recycling per Ton of PET Plastic Waste

Revenues per ton of plastic waste [$/t]	146.94
Total costs per ton of plastic waste [$/t]	157.52
Profit per ton of PET plastic waste [$/t]	**−10.58**

Since the profit is negative, recycling of plastic waste is not absolutely profitable. For this to have an absolute profit, the break-even price of recycled PET would need to be **$1.37**/kg instead of $1.28/kg.

This profitability depends significantly on the consumers. If plastic waste is not disposed of in the recycling trash can, plastic waste will not end up in the MRF, and so it will not be reprocessed at all. In this analysis, it was assumed that all waste was disposed of properly.

4.4.5 Influence of Oil Price on Profitability of Plastics Recycling

Oil is the most important raw material for plastics. As already mentioned, 1 kg of plastic requires about 2 kg of crude oil (including processing and raw materials for plastics). For this reason, the oil price has a great influence on the plastic price and the profitability of the whole plastic recycling process. The lower the oil price, the lower the price of the recycled plastic, the less profitable the recycling process becomes. Furthermore, a lower oil price reduces the costs of producing new, or virgin, plastic material, which is another challenge for plastics recycling.

In 2015 and 2016, low global oil prices significantly reduced the profitability of plastics recycling in the United States. In Newark, for example, the value for 1 bale of recycled plastics decreased from \$230 to \$112. One of the consequences was that Infinitus Energy, which had just opened a \$35 million recycling center in Montgomery, Alabama, in 2014, shut that facility down in October 2014 since it was incurring losses only. [54]

Figure 4.4 shows the changing price of 1 t of regrind PET compared to 1 barrel (158.9873 L) of oil. Since the original prices of the plastics were in Euros, the rate of 1 Dollar per Euro was taken from X-Rates. For a better comparison of different years, prices were inflation-adjusted based on the prices of October 2008 (both PET and oil). [55, 56, 57, 58]

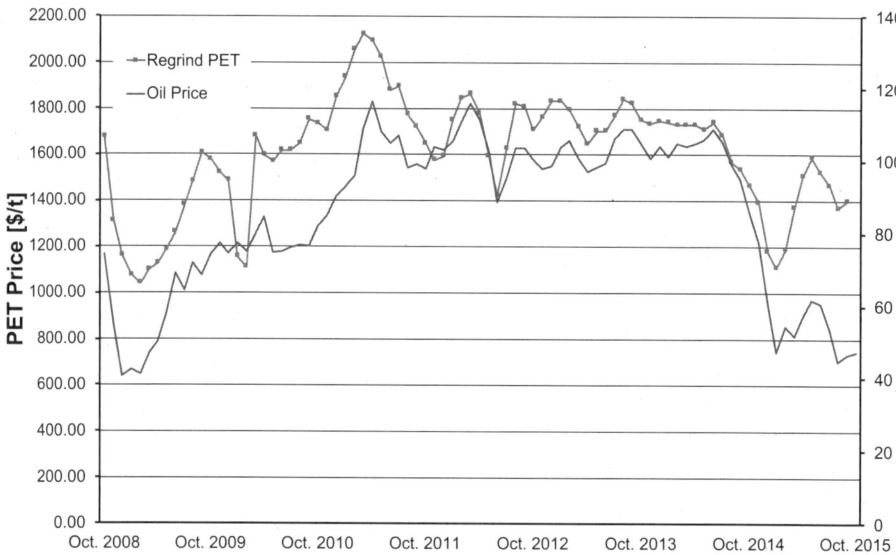

Figure 4.4 Oil (1 barrel (bbl) = 158.9873 L) and PET (1 t regrind) prices between October 2008 and October 2015 (inflation-adjusted)

Figure 4.4 demonstrates that the PET price is following the oil price changes, since the peaks of the regrind PET price graph line are lagging behind the peaks of the oil barrel price graph line by approximately 1 month. This can especially be observed in November 2009, October 2010, November 2014, and August 2015.

To analyze price correlations between PET and oil, a helpful tool is the calculation of the *Pearson product-moment correlation coefficient* $r_{x,y}$, which is a measure of degree of linear dependence between two variables. The calculation for $r_{x,y}$ is given in Equation 4.1.

$$r_{x,y} = \frac{\mathrm{cov}(x,y)}{\sigma_x \sigma_y} = \frac{\sum_{k=1}^{n}(x_k - \bar{x})(y_k - \bar{y})}{\sqrt{\left(\sum_{k=1}^{n}(x_k - \bar{x})^{?}\right)\left(\sum_{k=1}^{n}(y_k - y)^{?}\right)}} \qquad 4.1$$

The correlation coefficient ranges from $r_{x,y} \in [-1,1]$. A correlation coefficient $r_{x,y}$ of 1 implies a perfect relationship between x and y (a positive correlation). A value of −1 implies a perfect opposing relationship between the x and y (a negative correlation). A value of 0 implies no correlation between the variables. [59] The correlation between the two prices is shown graphically in Figure 4.5.

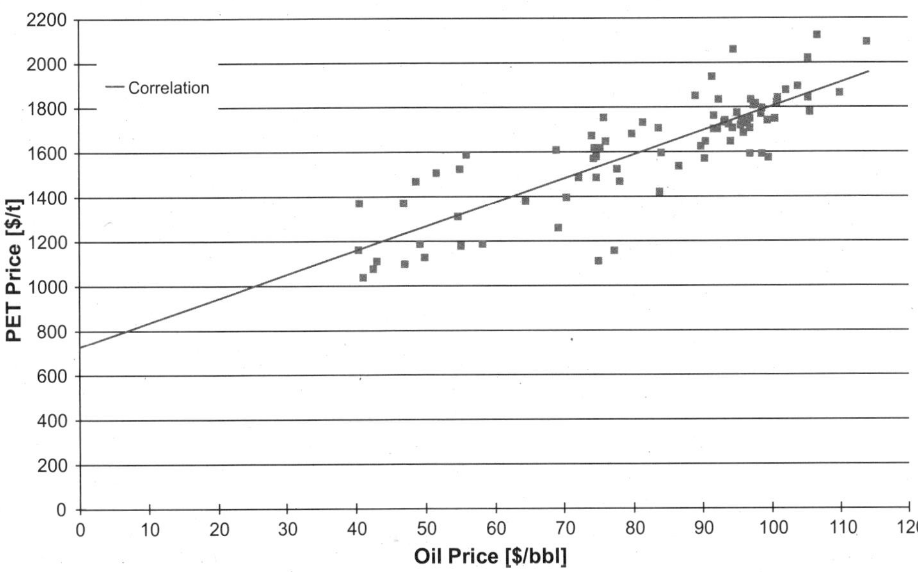

Figure 4.5 Correlation between PET price (per ton) and oil price (per barrel). bbl, barrel (158.9873 L)

For inflation-adjusted prices, the correlation coefficient is 0.84. This value reflects the results shown in Figure 4.4 and Figure 4.5: prices of PET and oil have a positive but not perfectly linear correlation.

Based on the correlation curve and knowing that the break-even price of recycled PET is $1.37 per kilogram, or $1,370.00 per ton, the required oil price for profitability is about $60.00 per barrel, as presented in Figure 4.6. In March 2020, the oil price was around $45 per barrel, which is too low to make plastic recycling profitable.

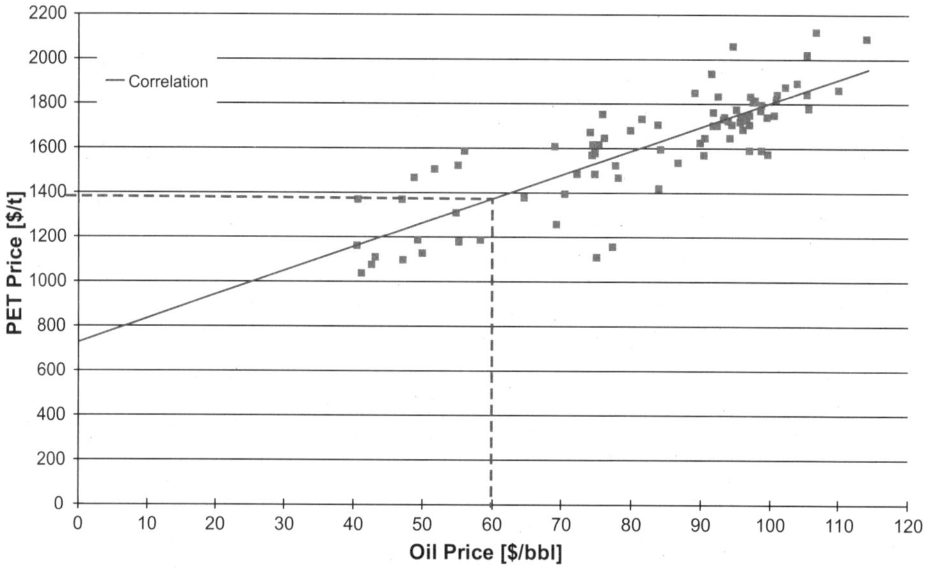

Figure 4.6 Required oil price per barrel for break-even recycled PET price per ton

■ 4.5 Influence of China's Import Ban on Profitability of Plastics Recycling

The recycled plastics are only partially reused in the U.S. By 2017, more than 60% of recycled plastic waste was *exported*, as shown in Figure 4.7. [61]

Between 2017 and 2018 a significant shift in exported waste between countries can be observed. The import ban on plastic waste in China, which is discussed in more detail in Section 7.2, has reduced U.S. exports to China by almost 90%. Despite the increase in exports to other countries, especially Malaysia, Thailand, and Vietnam, approximately one third less plastic waste was exported in the first half of 2018 compared to the first half of 2017. This poses a major challenge for the U.S., as there were approximately 600,000 tons of plastic waste "left over" in 2018. For the recycling industry, a significant reduction in the selling price could be recorded: Between June 2018 and June 2019, the price for 1 kg of recycled PET pellets decreased from \$1.59 to \$1.37. This further reduces the economic viability of recycling in the U.S. Accordingly, new recycling options for plastic waste must be developed or plastic waste generally avoided. [62]

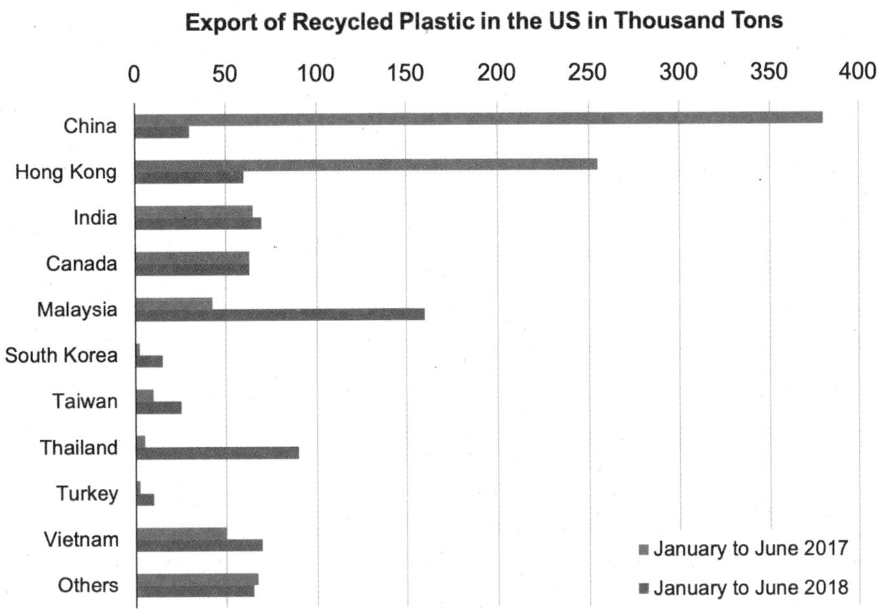

Figure 4.7 Distribution of exports of U. S. plastic waste by country

4.6 Conclusion: Economical Feasibility of Plastics Recycling

In the economic analysis presented using the static profit comparison method, the profit of handling plastic waste was analyzed for three different procedures. The profits of the three waste handling procedures for 1 t of waste are summarized in Figure 4.8. These numbers presume an ideal handling of plastic waste, which means all waste is disposed of properly, so that the plastic waste composition conforms with the plastic waste composition found in the MSW in 2017 (Table 2.1).

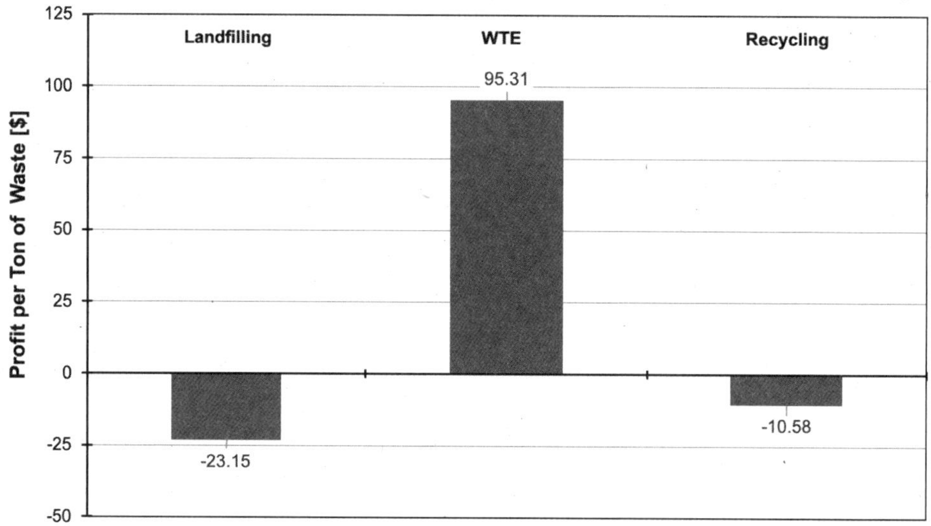

Figure 4.8 Profitability of recycling 1 t of plastic for three different waste handling procedures

Incineration with energy recovery (waste-to-energy) is the only handling method with absolute profitability (profit > $0.00) and therefore relative profitable as well. However, this is only valid in the case of incineration of plastics only and no other materials, which is really a theoretical assumption.

As already mentioned during this analysis, the numbers depend significantly on the size and yearly capacity of the facilities. The bigger the facility, the lower the costs and the higher the profit.

In the future, costs for landfilling will probably increase due to a lack of land available, so its profitability will further decrease.

The profitability of plastic recycling depends on two factors, which cannot be influenced by the factory: (1) the oil price and (2) the plastic recycling ratio of consumers, which depends on whether consumers recycle their waste properly, so that all waste plastic gets to the recovery factories to be recycled.

References

[1] BWL Institut Basel. *Investitionsrechnung*. Basel, Switzerland. 2013.

[2] Shim, J. K., and Siegel, J. G. *Budgeting Basics and Beyond,* 3rd ed. John Wiley & Sons, Hoboken, NJ. 2009.

[3] Götze, U., Northcott, D., and Schuster, P. *Investment Appraisal: Methods and Models,* 2nd ed. Springer-Verlag, Berlin, Germany. 2015.

[4] Enseling, A. *Leitfaden zur Beurteilung der Wirtschaftlichkeit von Energiesparinvestitionen im Gebäudebestand.* Access Date: 2016/02/23. Available: *http://www.iwu.de/fileadmin/user_upload/dateien/energie/klima_altbau/leitfaden_wirtschaftlichkeit.pdf*

[5] Bradu, M. Statistical-financial valuation methods of the investment projects. *Theoretical and Applied Economics Review.* vol. 508, pp. 49–52, 2007.

[6] Hering, E. *Investitions- und Wirtschaftlichkeitsrechnung für Ingenieure.* Springer Fachmedien Wiesbaden, Wiesbaden, Germany. 2014.

[7] Leeman, R. *Methoden der Wirtschaftlichkeitsanalyse von Energiesystemen.* Bundesamt für Konjunkturfragen (BfK), Impulsprogramme RAVEL, Bern, Switzerland. 1992.

[8] United States Senate. *Solid Waste Disposal Act (RCRA).* Washington, D.C. 2002.

[9] United States Environmental Protection Agency (EPA). *Regulatory impact analysis for the proposed revisions to the emission guidelines for existing sources and supplemental proposed new source performance standards in the municipal solid waste landfills sector.* Washington, D.C. 2015.

[10] United States Environmental Protection Agency (EPA). *Municipal Solid Waste Landfills–Economic Impact Analysis for the Proposed New Subpart to the New Source Performance Standards.* Washington, D.C. 2014.

[11] Duffy, D.P. Landfill Economics, Part II: Getting down to business. *MSW Management.* July/August 2005.

[12] Duffy, D.P. Landfill Economics, Part III: Closing up shop. *MSW Management.* September/October 2005.

[13] Duffy, D.P. Landfill Economics, Part I: Siting. *MSW Management.* vol. 15, issue 3, p. 118, May/June 2005.

[14] Caterpillar. *Specalog for 950H/962H–Wheel Loaders.* 2016.

[15] Caterpillar. *Specalog for 826G Series II–Landfill Compactor.* 2016.

[16] Caterpillar. *Specalog for D7R–Track-Type Tractor.* 2016.

[17] Eilrich, F.E., Doeksen G.A., and Van Fleet, H. *An Economic Analysis of Landfill Costs to Demonstrate the Economies of Size and Determine the Feasibility of a Community Owned Landfill in Rural Oklahoma.* Paper presented at the Southern Agricultural Economics Association, Mobile, AL, February 1-5, 2003.

[18] Glassdoor. *Waste Management Scale Attendant Salaries.* Access Date: 2016/04/13. Available: *https://www.glassdoor.com/Salary/Waste-Management-SCALE-ATTENDANT-Salaries-E2094_D_KO17,32.htm*

[19] Salary.com. *Facility Manager Salaries.* Access Date: 2016/04/13. Available: *http://www1.salary.com/Facilities-Manager-Salary.html*

[20] Salary.com. *Heavy Equipment Operator Salaries.* Access Date: 2016/04/13. Available: *http://www1.salary.com/Heavy-Equipment-Operator-Salaries.html*

[21] Salary.com. *General Laborer Salaries.* Access Date: 2016/04/13. Available: *http://www1.salary.com/General-Laborer-Salary.html*

[22] National Solid Wastes Management Association. *Modern Landfills: A Far Cry from the Past.* Washington, D.C. 2008.

[23] Gang, D.W. USA Today. *5 years after coal-ash spill, little has changed.* Access Date: 2016/04/13. Available: *http://www.usatoday.com/story/news/nation/2013/12/22/coal-ash-spill/4143995/*

[24] Roland Berger Strategy Consultants. Trend 3: scarcity of resources. In: *Roland Berger Trend Compendium 2030.* 2015.

[25] National Solid Wastes Management Association. *Municipal Solid Waste Landfill Facts.* Washington, D.C. 2012.

[26] Michaels, T. *The 2014 ERC Directory of Waste-to-Energy Facilities.* Energy Recovery Council, Arlington, VA. 2014.

[27] Cekirge, H.M., Ouda, O.K. M., and Elhassan, A. Economic analysis of solid waste treatment plants using pyrolysis. *American Journal of Energy Engineering.* vol. 3, pp. 11-15, 2015.

[28] ICF International. *Economic Analysis of New Waste-to-Energy Facility in Metro Vancouver.* Paper prepared for Belkorp Environmental Services, Inc. 2014.

[29] Schneider, D.R., Lončar, D., and Bogdan, Ž. Cost analysis of waste-to-energy plant. *Strojarstvo.* vol. 52, pp. 369 -378, 2010.

[30] USforex. *Yearly Average Rates.* Access Date: 2016/04/13. Available: *http://www.usforex.com/forex-tools/historical-rate-tools/yearly-average-rates*

[31] The World Bank. *Price level ratio of PPP conversion factor (GDP) to market exchange rate.* Access Date: 2016/04/13. Available: *http://data.worldbank.org/indicator/PA.NUS.PPPC.RF*

[32] Glassdoor. *Maintenance Engineer Salaries.* Access Date: 2016/04/13. Available: *https://www.glassdoor.com/Salaries/maintenance-engineer-salary-SRCH_KO0,20.htm*

[33] Salary.com. *Energy Manager Salaries.* Access Date: 2016/04/13. Available: *http://www1.salary.com/Energy-Manager-Salaries.html*

[34] Salary.com. *Environmental Engineer III Salaries.* Access Date: 2016/04/13. Available: *http://www1.salary.com/Environmental-Engineer-III-Salaries.html*

[35] U.S. Energy Information Administration. *Average Price of Electricity to Ultimate Customers by End-Use Sector.* Access Date: 2020/02/15. Available: *https://www.eia.gov/electricity/monthly/epm_table_grapher.php?t=epmt_5_6_a*

[36] Themelis, N.J., and Mussche, C. *2014 Energy and Economic Value of Municipal Solid Waste (MSW), Including Non-Recycled Plastics (NRP) Currently Landfilled in the Fifty States.* Earth Engineering Center, Columbia University, New York, NY. 2014. pp. 1–40.

[37] Energy Information Administration, Office of Coal, Nuclear, Electric and Alternate Fuels. *Methodology for allocating municipal solid waste to biogenic and non-biogenic energy.* U.S. Department of Energy, Washington, D.C. 2007.

[38] Rand, T., Haukohl, J., and Marxen, U. *Municipal Solid Waste Incineration: A Decision Makers' Guide.* The World Bank, Washington, D.C. 2000.

[39] Davis, J., Haase, S., and Warren, A. *Waste-to-Energy Evaluation: U.S. Virgin Islands.* National Renewable Energy Laboratory (NREL), Golden, Colorado. 2011.

[40] Arsova, L., Van Haaren, R., Goldstein, N., Kaufman, S.M., and Themelis, N.J. The state of garbage in America. *BioCycle.* vol. 49, p. 22, 2008.

[41] Van Haaren, R., Themelis, N., and Goldstein, N. The state of garbage in America. *BioCycle.* vol. 51, pp. 16–23, 2010.

[42] Psomopoulos, C.S., Bourka, A., and Themelis, N.J. Waste-to-energy: a review of the status and benefits in USA. *Waste Management.* vol. 29, pp. 1718–1724, 2009.

[43] University of Cambridge and University of Cambridge-MIT Institute. *The ImpEE Project: Recycling of Plastics.* Department of Engineering, University of Cambridge, Cambridge, UK. 2005.

[44] Kessler Consulting Inc. *MRFing Our Way to Diversion: Capturing the Commercial Waste Stream.* Tampa, FL. 2009.

[45] Pellitteri Waste Systems. *Recycling Center Video - Watch your Recycling get sorted.* Access Date: 2016/04/13. Available: *http://www.pellitteri.com/news.jsp?id=1716*

[46] Burns & McDonnell. *Mixed waste processing economic and policy study.* Austin, TX. Prepared for American Forest & Paper Association. 2015.

[47] Pressley, P.N., Levis, J.W., Damgaard, A., Barlaz, M.A., and DeCarolis, J.F. Analysis of material recovery facilities for use in life-cycle assessment. *Waste Management.* vol. 35, pp. 307–317, 2015.

[48] Gershman, Brickner & Bratton, Inc. *Materials Recovery Facility (MRF) Feasibility Report.* Fairfax, VA. Prepared for City of Tucson Environmental Services. 2008.

[49] Axion Consulting. *A Financial Assessment of Recycling Mixed Plastics in the UK: Financial modelling and assessment of mixed plastic separation and reprocessing (WRAP project MDP021).* Prepared for Waste Recovery Action Programme (WRAP). 2009.

[50] Al-Salem, S., Lettieri, P., and Baeyens, J. Recycling and recovery routes of plastic solid waste (PSW): a review. *Waste Management.* vol. 29, pp. 2625–2643, 2009.

[51] X-Rates. *Currency-calculator: British Pound to Dollar.* Access Date: 2016/04/26. Available: *http://www.x-rates.com/calculator/?from=GBP&to=USD&amount=1*

[52] Plasticnews. *Plastic Resin Pricing – Recycled Plastics.* Access Date: 2020/03/15. Available: *https://www.plasticsnews.com/resin/currentPricing/recycled-plastics*

[53] United States Environmental Protection Agency (EPA). *Advancing Sustainable Materials Management: Facts and Figures 2013. Assessing Trends in Material Generation, Recycling and Disposal in the United States.* Washington, D. C. 2015.

[54] Gelles, D. *Losing a Profit Motive.* New York Edition: New York Times. February 12, 2016. p. B5. Available: *https://www.nytimes.com/2016/02/13/business/energy-environment/skid-in-oil-prices-pulls-the-recycling-industry-down-with-it.html?_r=0*

[55] EUROINVESTOR. *Oil price development.* Access Date: 2016/04/13. Available: *http://www.euroinvestor.com/exchanges/gtis-energy/brent-oil/2327059*

[56] Plasticker. *The Home of Plastics: Raw Materials & Prices.* Access Date: 2016/04/13. Available: *http://plasticker.de/preise/preise_monat_multi_en.php*

[57] X-Rates. *US Dollar per 1 Euro Monthly average.* Access Date: 2016/04/13. Available: *http://www.x-rates.com/average/?from=EUR&to=USD&amount=1&year=2008*

[58] US Inflation Calculator. *Historical Inflation Rates: 1914–2016.* Access Date: 2016/04/13. Available: *http://www.usinflationcalculator.com/inflation/historical-inflation-rates/*

[59] Runkler, T. A. *Korrelation.* In: Data Mining: Methoden und Algorithmen intelligenter Datenanalyse. Vieweg+Teubner, Wiesbaden, Germany. 2010. pp. 55–63.

[60] Karidis, A. *Supply and Demands Drives Rising Tip Fees.* Access Date: 2020/03/15. Available: *https://www.waste360.com/landfill-operations/supply-and-demand-drives-rising-tip-fees*

[61] Clarke, J. S., and Howard, E. *US Plastic Waste Exports to Developing Countries, Causing Environmental Problems at Home and Abroad.* Access Date: 2020/03/15. Available: *https://unearthed.greenpeace.org/2018/10/05/plastic-waste-china-ban-united-states-america/*

[62] Dell, J. *157,000 Shipping Containers of U. S. Plastic Waste Exported to Countries with Poor Waste Management in 2018.* Access Date: 20/03/15. Available: *https://www.plasticpollutioncoalition.org/blog/2019/3/6/157000-shipping-containers-of-us-plastic-waste-exported-to-countries-with-poor-waste-management-in-2018*

5 Environmental Analysis of Plastic Waste Handling

Chapter 4 showed that incineration with energy recovery is economically the best method of plastic waste handling. However, to improve waste handling and address the challenges resulting from the continuous increase of plastic waste mentioned in Chapter 2, plastic waste handling methods have to be analyzed from an *environmental point of view* as well. For present and especially future society, environmental aspects have a much higher priority than economic ones: at least 8 million tons of plastics are leaked into the ocean each year, which is equivalent to dumping the contents of one garbage truck into the ocean every minute. If the plastic waste handling does not change, the leakage is expected to increase to two trucks per minute by 2030 and even four per minute by 2050. There is only one planet and this planet needs to be protected. An environmentally sensitive handling of plastic waste is critical to the protection of the environment. [1]

As already mentioned in Section 1.2, recycling is the preferred method of waste handling by the Environmental Protection Agency (EPA), but only 8.4% of the total plastic waste is recycled (Table 2.1). Chapter 5 will explain why from an environmental point of view, compared with landfilling and incineration, recycling is the most favorable option.

5.1 Environmental Analysis of Landfilling

The percentage of plastic waste in landfills is increasing compared to its percentage in the total MSW because of the low percentage of plastic waste recycled (8.4%) and incinerated (16.4%). Especially the recycling rate of plastics is significantly lower than the recycling rate of MSW in total (35.2%). Due to their current waste treatment, plastics make up 19.2% of the total waste in landfills, which is significantly more than the 13.2% they make up of total MSW. Only food scraps account for a larger percentage of waste in landfills (22.0%). Waste disposal of plastic in landfills has several negative effects on the environment. [2]

First of all, disposing of plastic waste in landfills is equivalent to the loss of scarce resources. As already mentioned in Chapter 2, producing 1 kg of plastic requires 2 kg of crude oil, which is lost in cases where plastics are disposed of in landfills and not recycled. Furthermore, landfills occupy land, which is becoming a more and more scarce resource as well. The 1,269 landfills in the United States covered an area of around 120,000 ha in 2017. Knowing that plastics make up 19.2% of the total waste in landfills means they occupy around 23,040 ha. According to a study by Yale University, the available space for landfills that meet the RCRA D regulations will only last for another 70 years, even if parts of these areas can be reused in some cases. Bearing in mind that plastic waste is increasing continuously and over 80% of it is placed in landfills, this scarcity of land is becoming a big ecological challenge for plastic waste handling. [3, 4, 5, 6, 7]

As a result of disposing of plastic in landfills, the occupied soil and its groundwater is polluted by *leachate*. Leachate describes liquid, whose primary source is precipitation, that passes through the landfilled waste and collects and dissolves chemicals from the refuse as it percolates. Even if the activity of landfills in general is limited to 30 years and RCRA D regulates the aftercare of landfills, leachates cannot be completely prevented and make a revival of the land nearly impossible. [8, 9, 10]

Another way plastics in landfills are polluting the environment is via greenhouse gas (GHG) production. Since plastics normally do not contain biodegradable carbon, they do not generate methane gas (CH_4) and are not considered as storing any carbon when landfilled. The only emissions associated with landfilling plastics are from transportation to the landfill and moving waste into the landfill. But the relatively new, so-called "environmentally friendly" plastic bags, being biodegradable and taking about 3 years to break down into practically nothing, produce greenhouse gases. After being placed in a landfill, the organic plastic bags are initially decomposed by aerobic bacteria and afterwards broken down through fermentation into gases and short-chain organic compounds. These compounds form substrates for the growth of methanogen bacteria, which convert the fermentation products into biogas consisting of approximately 50% biogenic carbon dioxide (CO_2) and 50% CH_4 (methane).

At present, the amount of GHGs produced by these biodegradable plastics is low, especially compared to incineration (see Section 5.2). However, if the production of biodegradable plastics increases and its waste is disposed of in landfills, the GHG production will become a bigger issue of landfilling. [2, 11, 12]

The biggest environmental pollution caused by landfilling plastics is marine pollution. Through mismanagement in landfills, the United States releases between about 0.04 and 0.11 Mt of plastic waste into the ocean every year, and this is rising. *Microplastics* are especially a huge risk. Due to ultraviolet (UV) radiation, plastics in landfills start degrading and disintegrate into their different chemical components and additives. Both the big and small pieces of plastic pose a risk for marine animals or birds. Often they cannot distinguish between plastics and food and eat the plastic waste, which afterward stays in their bodies and causes intestinal obstructions. [13, 14]

5.2 Environmental Analysis of Incineration with Energy Recovery (Waste-to-Energy Facilities)

As it has been pointed out already, 16.4% of plastic waste is incinerated with energy recovery. WTE facilities burn waste, including plastic waste, to produce electricity (see Sections 1.2.2 and 4.3 for a detailed description). This incineration leads to an ecologically valuable reduction of plastic waste: the remaining, nonhazardous combustion ash amounts to about 25% of the weight of the unprocessed waste input and about 10% of its volume. This reduction of weight and especially volume is highly important for the environment, considering that land is a scarce resource (Section 5.1). Including landscaping and auxiliary buildings, a WTE plant, which lasts over 30 years, processes 1 Mt of waste each year on less than 10 ha of land. In total, this would be 30 Mt of waste on 10 ha of land per plant. In comparison, landfilling 30 Mt of waste would require 300 ha of land. Furthermore, a new WTE plant could be built on the site of the existing WTE plant. Landfill sites cannot be used for anything else again and new green fields would need to be converted to landfills. [2, 15, 16, 17, 18]

WTE facilities can also help to reduce fossil fuel use and foreign oil dependence. Operating continuously (24 hours a day, 7 days a week), they provide baseload electricity to communities. One ton of MSW has the chemical content of 0.4 t of oil: 1 t of nonrecycled plastic has the chemical content of 1.4 t of oil. [17, 19]

But combustion also induces negative ecological impacts. First of all, burning plastic means burning of valuable, scarce resources. As mentioned in Chapter 2 already, producing 1 kg of plastic requires 2 kg of oil. Furthermore, the ash from burning plastic waste, although its weight and volume are lower than the untreated MSW, ends up in landfills and therefore requires land. Most states have strict regulations to ensure that ash does not exhibit any toxicity characteristics, but the influence of plastic ash on the environment has not been completely researched yet. [15, 19]

Another controversial subject of waste-to-energy plants, more than any other plastic handling method, is noise. Trucks bringing solid waste to the facility, plant operations, and fans are sources of noise pollution, which affects human health significantly. [20]

The biggest issue of burning plastic is the generation of pollutants, especially CO_2. Since plastic is created from a fossil fuel, its combustion is considered an anthropogenic source of carbon emissions. An EPA study revealed that incinerators are the dirtiest electricity production option, releasing more greenhouse gases than coal-fired power stations per unit of energy generated.

Table 5.1 shows the *net emission factor* for combustion of 1 t of high-density polyethylene (HDPE), low-density polyethylene (LDPE), and polyethylene terephthalate (PET) in metric tons of a carbon dioxide equivalent ($MtCO_2E$)[1] calculated using the EPA's *Waste Reduction Model (WARM)*. This factor includes the emissions associated with transporting (903 km per shipment) the plastic waste to WTE facilities and emission savings associated with the avoided emissions of burning conventional fossil fuels for utilities. It shows that the production of greenhouse gases through waste combustion is much higher than the emission savings. [11, 21]

Table 5.1 Net Emissions Factor Due to Combustion for Various Plastics

Material	Transportation to Combustion [$MtCO_2E/t$]	CO_2 from Combustion [$MtCO_2E/t$]	Utility Emissions Avoided [$MtCO_2E/t$]	Net Emissions Factor [$MtCO_2E/t$]
HDPE	0.033	3.075	−1.664	1.444
LDPE	0.033	3.075	−1.664	1.444
PET	0.033	2.249	−0.871	1.411

CO_2, dioxins, and particles contribute to negative effects for the environment, such as climate change, smog, and acidification, and for the human body, such as asthma, lung damage, cardiac problems, and nervous system damage. [15, 22]

5.3 Environmental Analysis of Recycling

The recycling rate of plastic materials in 2017 was 8.4%, thus much lower than the recycling rate of general MSW (35.2%). Despite this low rate, plastic recycling has a big positive ecological impact: it provides opportunities to reduce quantities of waste requiring disposal, oil usage, and carbon dioxide emissions. [2, 23]

[1] $MtCO_2E$ (metric tons of carbon dioxide equivalent): This describes how much global warming a given type and amount of greenhouse gas causes using the equivalent of CO_2 as a reference.

Recycling of plastics means waste reduction. In 2017, the total plastic waste produced was 35.4 million tons. Even at the relatively low recycling rate of 8.4%, it means 2.96 million tons were neither landfilled nor burned, thus not polluting the environment. [2, 24]

Furthermore, plastic recycling is equivalent to the reuse of scarce resources, especially oil. Nowadays, plastics are almost completely derived from petrochemicals, which are produced from fossil oil and gas. Since manufacturing of plastics also requires energy, a similar additional quantity of fossil fuels is used for their production. Reprocessing plastics is consequently the same as reuse of this important resource. [23]

The key benefit of recycling plastic is *the reduction of required plastics production:* less production means less energy use, which simultaneously leads to the reduction of CO_2 and greenhouse gas emissions. Considering the difference between the energy use for producing virgin PET and HDPE and for reprocessing these products at the end of their life, recycling only these two plastics in the United States could save enough energy each year to power 750,000 homes. [25]

Table 5.2 Net Emissions Factor Due to Combustion and Energy Savings for Recycled versus Virgin Plastics

		HDPE	LDPE	PET
Virgin input [MtCO_2E/t]	Process energy	1.560	1.905	1.796
	Transportation energy	0.036	0.036	0.036
	Process non-energy	0.172	0.172	0.100
Recycled input [MtCO_2E/t]	Process energy	0.118	0.118	0.118
	Transportation energy	0.045	0.045	0.045
	Process non-energy	–	–	–
Savings by recycling [MtCO_2E/t]	Process energy	−1.442	−1.787	−1.678
	Transportation energy	0.009	0.009	0.009
	Process non-energy	−0.172	−0.172	−0.100
	Total savings	**−1.605**	**−1.950**	**−1.769**

Table 5.2 shows the difference between emissions from manufacturing 100% virgin material and 100% recycled material, calculated using the WARM method, which breaks down the emission into

- Process energy emissions
- Transportation emissions
- Process non-energy emissions

It can be seen that manufacturing of recycled HPDE, LDPE, and PET significantly reduces GHG emissions compared to producing the same amount of virgin material. Among these plastics, LDPE recycling shows the largest GHG emission savings. [11, 25]

Reduction of waste, energy use, and GHG emissions are several positive effects of recycling plastics on the environment. These positive ecological impacts are reflected in the EPA's waste management hierarchy, which superordinates recycling to incineration and landfilling of plastic waste. [2, 24]

5.4 Conclusion: Environmental Necessity of Plastics Recycling

Considering all waste handling options from an ecological point of view, it has been established that recycling clearly is the best way to handle plastic waste. Besides the reduction of waste, it leads to energy savings and decreased GHG emissions.

Recycling is not only a waste management strategy; it further implements the concept of industrial ecology, that there is no waste but only new products. [25]

On this account, the recycling process needs to be improved so that it is both ecologically and economically desirable. Therefore, Chapter 6 will consider two different ways of economically improving the plastics recycling process and making it even more indispensable from an ecological perspective.

References

[1] Project MainStream. *The New Plastics Economy: Rethinking the Future of Plastics.* Ellen MacArthur Foundation, The World Economic Forum, and McKinsey & Company. 2016.

[2] United States Environmental Protection Agency (EPA). *Advancing Sustainable Materials Management: 2016 and 2017 Tables and Figures.* Washington, D.C. 2019.

[3] Powell, J.T., Townsend, T.G., and Zimmermann, J.B. Estimates of solid waste disposal rates and reduction targets for landfill gas emissions. *Nature Climate Change.* vol. 6, pp. 162–166, 2016.

[4] Statista. *Number of municipal waste landfills in the U.S. from 1990 to 2017.* 2020. Available: *https://www.statista.com/statistics/193813/number-of-municipal-solid-waste-landfills-in-the-us-since-1990/*

[5] United States Senate. *Solid Waste Disposal Act (RCRA).* Washington, D.C. 2002.

[6] Themelis, N., and Ulloa, P. Capture and utilization of landfill gas. *Renewable Energy.* pp. 77–81, 2005.

[7] National Solid Wastes Management Association. *Modern Landfills: A Far Cry from the Past.* Washington, D.C. 2008.

[8] United States Environmental Protection Agency (EPA). *Regulatory Impact Analysis for the Proposed Revisions to the Emission Guidelines for Existing Sources and Supplemental Proposed New Source Performance Standards in the Municipal Solid Waste Landfills Sector.* Washington, D.C. 2015.

[9] Scharff, H., van Zomeren, A., and van der Sloten, H.A. Landfill sustainability and aftercare completion criteria. *Waste Management & Research.* vol. 29, pp. 30–40, 2011.

[10] Kjeldsen, P., Barlaz, M.A., Rooker, A.P., Baun, A., Ledin, A., and Christensen, T.H. Present and long-term composition of MSW landfill leachate: a review. *Critical Reviews in Environmental Science and Technology.* vol. 32, pp. 297–336, 2002.

[11] United States Environmental Protection Agency (EPA). *Plastics.* Washington, D.C. 2010.

[12] Jacobsen, S. *Plastic Bag Pollution.* 2005. Available: *http://dpw.lacounty.gov/epd/PlasticBags/Articles/Googobits_07-21-05.pdf*

[13] Jambeck, J.R., Geyer, R., Wilcox, C., Siegler, T.R., Perryman, M., Andrady, A., Narayan, R., and Law, K.L. Plastic waste inputs from land into the ocean. *Science.* vol. 347, pp. 768–771, 2015.

[14] Klammer, P., and Blage, J. Wenn Plastik Tötet. *Focus.* vol. 13, pp. 110–114, 2013.

[15] Center for Sustainable Systems. *Municipal Solid Waste.* University of Michigan, Ann Arbor, MI. 2015.

[16] Psomopoulos, C.S., Bourka, A., and Themelis, N.J. Waste-to-energy: a review of the status and benefits in USA. *Waste Management.* vol. 29, pp. 1718–1724, 2009.

[17] Themelis, N.J., and Mussche, C. *Energy and Economic Value of Municipal Solid Waste (MSW), Including Non-Recycled Plastics (NRP), Currently Landfilled in the Fifty States.* Earth Engineering Center, Columbia University. New York, NY. 2014. pp. 1–40.

[18] Schwartz, M., and Disney, C. *Art of the Deal: Building WTE in the 21st Century.* Paper presented at The 16th Annual North American Waste-to-Energy Conference (NAWTEC16), Philadelphia, PA, May 19–21, 2008.

[19] Helou, A.E., Tran, K., and Buncio, C. *Energy recovery from MSW in California: Needs and challenges.* Paper presented at The 18th Annual North American Waste-to-Energy Conference (NAWTEC18), Orlando, FL, May 11–13, 2010.

[20] Texas Comptroller of Public Accounts. Municipal Solid Waste Combustion. In: *The Energy Report 2008.* Austin, TX. 2008. pp. 257–267.

[21] Bell, L., and Bremmer, J. *Burning waste for energy: It doesn't stack-up.* National Toxics Network, Inc., Bangalow, Australia. 2014.

[22] United States Environmental Protection Agency (EPA). *Air Regulations for Municipal Waste Combustors.* Washington, D.C. 2011.

[23] Hopewell, J., Dvorak, R., and Kosior, E. Plastics recycling: challenges and opportunities. *Philosophical Transactions of the Royal Society of London B: Biological Sciences.* vol. 364, pp. 2115–2126, 2009.

[24] United States Environmental Protection Agency (EPA). *Municipal Solid Waste in the United States: 2017 Fact Sheet.* Washington, D.C. 2019.

[25] American Chemistry Council. The word on plastic recycling: momentum. *SPE Plastics Engineering.* pp. 48–50, February 2016.

6 Optimization of Plastics Recycling

Chapter 5 concluded that recycling is the best option for handling plastic waste from an environmental point of view and can significantly contribute to minimizing air, soil, and marine pollution.

But, as presented in Chapters 2, 3, and 4, there are two central issues with recycling: on the one hand, only 9% of plastic waste in the United States is recycled at the moment due to technical limitations (see Chapter 3) and, on the other hand, recycling is currently unprofitable from an economic point of view due to low oil prices (see Chapter 4). Recycling and selling 1 t of recycled plastic results in a loss of more than $10.

To improve both profitability and recycling rate, two process optimization possibilities are presented in this chapter.

6.1 Optimization I: Reduction of Sorting Processes

The first process optimization proposed is reducing the number of sorting processes. Therefore, the so-called *dual-stream recycling* would need to be implemented. Dual-stream recycling means that the plastic waste is directly separated by consumers in their households, which is similar to systems established in Europe (see Section 7.1). Consequently, the sorting process in the materials recovery facility (MRF) is not required anymore. The optimized process is shown in Figure 6.1. [1]

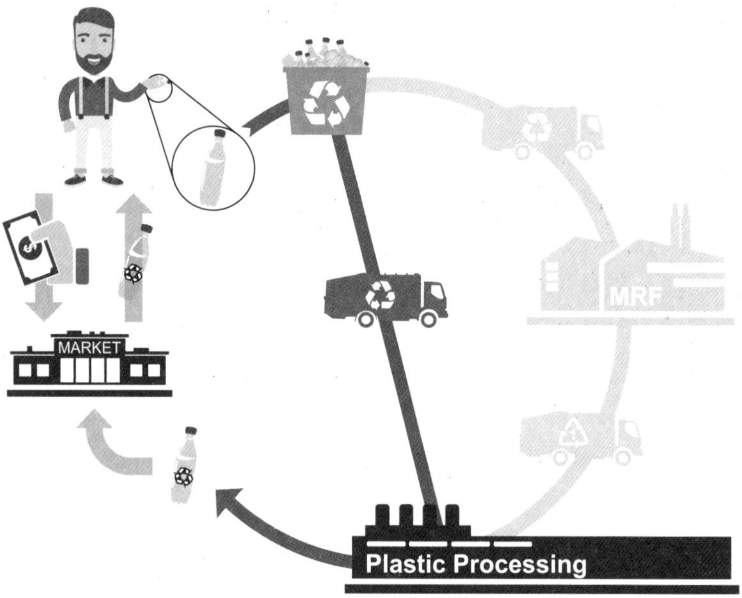

Figure 6.1 Optimization I: Dual-stream recycling

To calculate the profitability of the optimized process, the original profitability calculation of the plastic recycling process is used as a basis. The costs of polyethylene terephthalate (PET) processing as well as the revenues realized by selling recycled PET remain unchanged. Processing 1 t of plastic waste costs **$72.37** and the revenues for sale of 1 t of recycled plastic are **$146.94**. But to handle plastic in the same facility, additional machines and processes need to be installed. The additional costs are split up in two main categories: investment costs (1) and operation and maintenance costs (2). The assumptions for this optimization are shown in Table 6.1 and in more detail in Table 8.21 in the Appendix.

Table 6.1 Optimization I: Assumptions

Lifetime [years]	10
Yearly working hours [h]	6,240
Yearly plastic waste handling [t]	100,000
Total plastic waste capacity (10 years) [t]	1,000,000
Yearly PET capacity [t]	15,000
Total PET waste capacity (10 years) [t]	150,000
Separation efficiency [%]	91

Additional investment costs are split up in building and site, machine, and equipment costs. To handle plastic waste in only one facility, additional land, site work, and buildings as well as a scale house are required. These building and site costs amount to $1,775,000. Furthermore, three new machines need to be installed: a metering bin, an optical PET sorting machine, and a baler. The investment costs of all machines add up to $925,000. For additional conveyors, rolling stock, and waste collection cars, total costs are $1,250,000. As presented in Table 6.2, total additional investment costs are **$3,950,000** (see also Table 8.22 in the Appendix). [2, 3, 4]

Table 6.2 Optimization I: Additional Investment Costs

Additional building and site investment costs [$]	1,775,000
Additional machine investment costs [$]	925,000
Additional equipment investment costs [$]	1,250,000
Total additional investment costs [$]	**3,950,000**

Additional operating and maintenance costs are salaries of the additional personnel, operating and maintenance costs of the machines and the rolling stock, and especially transportation and collection costs. Yearly operating and maintenance costs are **$5,713,797**, so overall **$57,137,976**, as presented in Table 6.3 and in more detail in Table 8.23 in the Appendix. [3, 5, 6, 7, 8]

Table 6.3 Optimization I: Additional Operating and Maintenance (O&M) Costs

Personnel salaries per year [$]	963,000
Facility costs per year [$]	250,000
Machine O&M costs per year [$]	68,417
Rolling stock O&M costs per year [$]	748,380
Transportation and collection costs [$]	3,684,000
Yearly O&M costs [$]	**5,713,797**
Overall O&M costs (10 years) [$]	**57,137,976**

Summarizing both additional investment and operating and maintenance costs, total additional costs are **$61,087,976**. Since 100,000 t of plastic waste must be handled per year in this new facility area (to gain 15,000 t of PET waste, around 100,000 t of plastic waste has to be sorted), the additional costs of 1 t of plastic waste are **$61.09**.

Knowing that the revenues of recycling 1 t of plastic waste are **$146.94** and the costs for further processing the plastic waste are **$72.37**, the profitability of this optimization is calculated in Table 6.4.

Table 6.4 Total Profit per Ton of Plastics Recycled

Revenues per ton of plastics recycled [$/t]	146.94
Sorting [$]	61.09
PET processing [$]	72.37
Profit per ton of plastics recycled [$/t]	**13.48**

Table 6.4 shows that the profit of this optimization would be **$13.48** per ton of plastic waste recycled, so this process optimization has absolute profitability. Compared to the current plastic recycling procedure, the profit of the optimization is **$24.06** higher.

But this optimization has two risks, which cannot be included in the calculations. First, this optimized process depends even more on the **disposal behavior of consumers** than the current recycling process already does (see Chapter 2, motivation for recycling in different countries). If consumers not only have to distinguish between recyclables and non-recyclables but also between plastics, paper, glass, and metals, the risk of not properly disposing of plastic waste increases.

Second, the optimization has **hidden costs** for the transportation and collection of waste. The analysis only considered the recycling of plastic waste. In daily life, paper, glass, and metals should be recycled as well. If all of these materials were collected individually, transportation and collection costs would increase immensely.

6.2 Optimization II: Upcycling of Plastic Waste by Blending PP and LDPE

The second optimization possibility considered in this book includes the *extension of the analyzed PET reprocessing* to recycling polypropylene (PP) and low-density polyethylene (LDPE) by blending them into a compound. The process is presented in Figure 6.2.

Due to its desirable physical properties such as high tensile strength, high stiffness, and high chemical resistance, PP has been widely used as a packaging material. However, it shows poor impact strength at low temperatures and is susceptible to environmental *stress cracking*. LDPE is mostly used for bags and packaging films. Owing to its low mechanical properties but ease of processing, it is recycled and used for garbage bags. Therefore, blends of LDPE and PP have become a subject of great economic interest to improve the processing and mechanical properties of PP. [9]

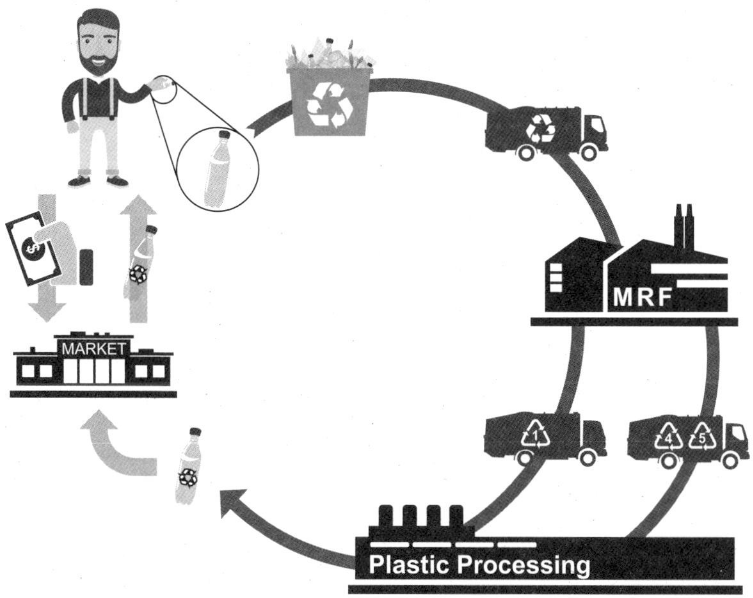

Figure 6.2 Optimization II: Extension of the process by blending recycled polypropylene (PP) and low-density polyethylene (LDPE)

Combinations of LDPE and PP are frequently found in polymer waste streams. But since their densities are very similar (PP: 946 kg/m^3, LDPE: 940 kg/m^3), they cannot be easily separated from each other by conventional sorting methods. Another motivation for blending recycled PP-LDPE is the high impact it has as shown in Table 2.1. PP and LDPE accounted for almost 50% of the total plastic waste in 2017. Not recycling these two plastics means wastage of important resources. However, it needs to be determined if blending recycled PP and LDPE would be economically profitable and as a result would optimize the plastic recycling process. [10, 11, 12]

Material tests showed that recycled PP and LDPE blends have material properties similar to virgin PP-LDPE blends. The effects of processing on the properties of recycled PP, recycled LDPE, and their blends were investigated using *melt flow index (MFI) measurements* (see Section 3.4 Plastic Degradation Mechanisms). These MFI measurements showed an increase with increasing amounts of rLDPE in blends, thus a decrease of the viscosity with increasing amounts of rLDPE, which is due mainly to the higher degradation of rLDPE, as shown in Figure 6.3.

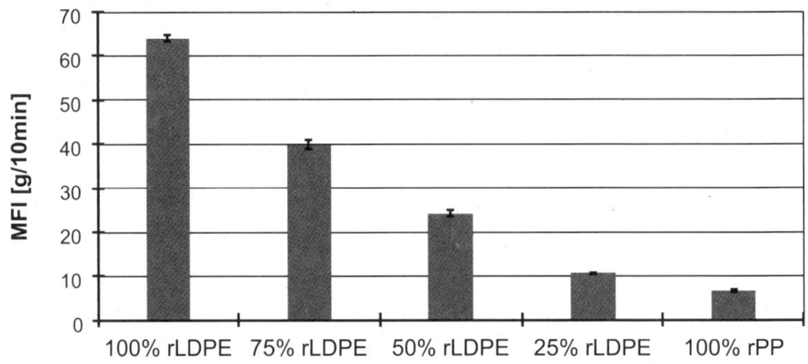

Figure 6.3 Melt flow index (melt) of recycled low-density polyethylene (rLDPE), recycled polypropylene (rPET), and their blends

These results also correspond to the mechanical tests. The higher the amount of rLDPE, the higher the maximum tensile strain but the lower the maximum tensile stress (Figure 6.4).

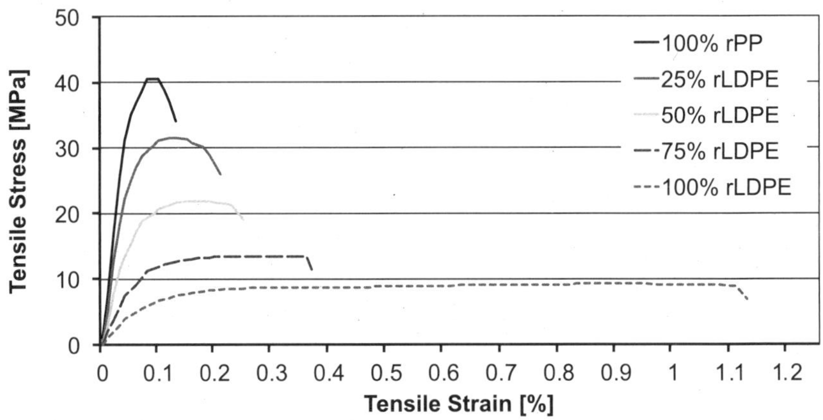

Figure 6.4 Tensile stress and tensile strain of recycled low-density polyethylene (rLDPE), recycled polypropylene (rPP), and their blends

Accordingly, blending recycled PP and LDPE can improve material properties to some extent. To calculate the profitability of optimization II, additional costs are calculated first. Afterwards, the revenues from selling rLDPE-rPP blends are calculated. From these, the profit of the optimization can be computed.

6.2.1 Additional Costs of LDPE–PP Recycling

To calculate the costs of optimization II, the plastic recycling process in Section 4.4 serves as a basis. The important assumptions are presented in Table 4.13, Table 4.17, and Table 6.5.

Table 6.5 Optimization II: Assumptions

Lifetime [years]	10
Percentage of PET in plastic waste [%]	14.39
Percentage of LDPE–PP in plastic waste [%]	45.70
Average price of recycled PET pellets [$/kg]	1.26
Average price of recycled LDPE–PP pellets [$/kg]	0.66
Total waste capacity (10 years) [t]	150,000
Separation efficiency [%]	91

First, the MRF process is extended by machines that are required to sort LDPE and PP. Since their densities are nearly the same and the optical sorter uses material densities to distinguish between different plastics, only one new sorter is required. Total investment as well as operating and maintenance costs are **$1,170,892**, or **$0.98** per t of plastic, as shown in Table 6.6 and in more detail in Table 8.24 in the Appendix. [2]

Table 6.6 Optimization II: Additional Costs of Low-Density Polyethylene and Polypropylene (LDPE–PP) Blending for Materials Recovery Facility (MRF)

Additional investment costs [$]	900,000
Additional O&M costs [$]	265,734
Overall additional MRF costs [$]	**1,170,892**
Additional MRF costs per ton [$/t]	**0.98**

Second, the plastic waste process needs to be adapted. Since plastics are sorted in the MRF first and transported separately to the plastic processing factory, PET and LDPE–PP are processed independently from each other. That means that a second metering bin, bale breaker, and washing station are required, as well as an LDPE–PP extruder. The yearly capacity for LDPE–PP is the same as for PET. Thus, investment costs for the metering bin, bale breaker, and washing station are the same as for PET. The LDPE–PP extruder is a bit cheaper than the PET extruder. Total machine investment costs of plastic processing are $24,069,560 (conveyers and installation included). In addition, facility space and rolling stock need to be expanded. These investments result in total costs of **$26,069,560** (Table 6.7 and Table 8.25 in the Appendix). [2, 13]

Table 6.7 Optimization II: Additional Investment Costs for Plastic Processing

Building and site investment costs [$]	1,950,000
Machine investments costs [$]	24,069,560
Additional equipment investment costs [$]	50,000
Total investment costs [$]	**26,069,560**

Since LDPE-PP requires an individual process, operation and maintenance costs are also similar to PET processing. All materials are processed in the same facility; therefore, no additional plant manager, marketing manager, and maintenance engineer is required. Another big difference between PET and LDPE-PP processing is electricity consumption of the extrusion process: LDPE-PP has a lower melting temperature than PET and consequently a lower electricity consumption. Yearly operation and maintenance costs of LDPE-PP processing are **$3,792,563**, thus **$37,925,630** after 10 years (Table 6.8 and Table 8.26 in the Appendix). [2, 13]

Table 6.8 Optimization II: Additional Operating and Maintenance (O&M) Costs of Plastic Processing

Personnel salaries per year [$]	1,500,000
Machine maintenance costs per year [$]	1,203,478
Machine operating costs per year [$]	977,578
Rolling stock operating costs per year [$]	111,507
Yearly additional O&M costs [$]	**3,792,563**
Overall additional O&M costs (10 years) [$]	**37,925,630**

Summarizing all additional investment and operating and maintenance costs over the factory's lifetime, total additional costs of the optimization process are $63,995,190, or $426.63 per t of LDPE-PP waste. Assuming a combined LDPE-PP content of 45.70% in the plastic waste (Table 2.1), additional processing costs per 1 t of LDPE-PP plastic are $193.95 (Table 6.9).

Table 6.9 Optimization II: Additional Costs per Ton for Plastic Processing

Investments costs [$]	26,069,560
Operating and maintenance costs [$]	37,925,630
Total costs (10 years – 150,000 t) [$]	**63,995,190**
Costs per ton of LDPE-PP [$/t]	**426.63**
Costs per ton of plastic waste (45.46%) [$/t]	**193.95**

6.2.2 Additional Revenues of LDPE–PP Recycling

To calculate the revenues of the LDPE–PP recycling process, the main task is to determine the selling price of rLDPE–rPP blends.

The results of the material analysis agreed with the behavior of virgin material: a higher amount of PP in blends leads to increased stiffness and strength of the material, which are desirable physical properties for packaging material. However, a greater portion of LDPE in the blends increased the maximum strain of material. Depending on the choice of application for the material, the ability of the material to elongate can also be an important criterion.

This shows that the value of a product is regulated by its demand. Determining an explicit selling price based on material properties is, also due to a lack of experience in this area, relatively arbitrary.

Since the ratio between LPDE (22.84% of plastic waste) and PP (22.62% of plastic waste) in the plastic waste is nearly identical, a ratio of 50% LDPE and 50% PP in the blend is assumed. For this reason, the price of recycled LDPE–PP pellets used for the following calculations is \$0.66 per kg, which is the mean value of rLDPE (\$0.52 per kg) and rPP (\$0.80 per kg) for March 2020. [14]

Calculating the revenues of plastic recycling, it needs to be remarked again that only 455 kg of 1 t of plastic waste is LDPE–PP (Table 2.1). The efficiencies of both the optical sorting at the MRF and the manual sorting during LDPE–PP processing need to be included as well. With an average price of \$0.66 per kg, additional revenues through optimization II are **\$236.53** (Table 6.10).

Table 6.10 Optimization II: Additional Revenues

Amount of LDPE–PP in plastic waste [kg]	454.60
Separation efficiency of MRF process [%]	86.45
Separation efficiency of PET processing [%]	91.00
Price of recycled LDPE–PP [\$/kg]	0.66
Total revenues from recycling LDPE–PP [\$]	**236.53**

6.2.3 Total Profit of Optimization II

Totaling revenues and costs from both the PET process (Section 4.4) and optimization II and subtracting the costs from the revenues, the profit from recycling 1 t of plastic waste is **\$31.65** (Table 6.11).

Table 6.11 Optimization II: Total Profit per Ton of Plastics Recycled

Revenues per ton of plastics [$/t]	383.47
Costs per ton of plastics [$/t]	351.82
Profit per ton of plastics recycled [$/t]	**31.65**

With optimization II, the profit would increase by **$42.50** per ton and make plastic recycling absolutely profitable. The implementation of an LDPE–PP process is relatively expensive, but since the additional costs of the MRF are very low, the profit increases significantly.

But as already mentioned earlier in this book, this is an ideal case with the prerequisite that all plastic is sorted and disposed of properly by consumers. Furthermore, the selling price of the rLDPE–rPP is of great importance for the revenue of the optimization process, and thus the profitability of the whole plastics recycling process. The **break-even selling price of rLDPE–rPP is $0.57/kg**. If the selling price for the blend were lower, the optimization would not be profitable. Furthermore, as analyzed in Section 4.4.5, the price of oil is an important factor for optimization II as well. The dependence of rLDPE and rPP prices on the price of oil is the same as for rPET. Assuming the oil prices of April 2016, optimization II turns plastic recycling into a profitable process.

6.3 Optimization III: Increasing the Recycling Rate

As presented in Section 6.2, upcycling[1] of plastic waste by blending several polymers improves the recycling process from both an economic and ecological point of view. In an optimal scenario, the recycling rate of plastic waste in the United States could be increased by 45.46% based on the numbers of 2017 (see Table 2.1).

However, the problem of this upcycling scenario is that the blended plastics become number 7 plastics (mixed plastic) and are therefore difficult to recycle a second or even a third time. Even if blending can improve the material properties, it is still desirable from an ecological perspective to recycle all plastics separately. Thus, multiple reprocessing cycles could be executed.

As mentioned in Section 6.2, separating PP (density: 946 kg/m^3) and LDPE (density: 940 kg/m^3) is nearly impossible by conventional and economically efficient sorting methods. However, based on the necessity of recycling discussed in Chapter 5, research regarding the separation of each plastic type has intensified. One

[1] Upcycling is the process of transforming waste materials into new materials or products of better quality or for better environmental value.

organization in the United States trying to increase the recycling rate of plastics is the *Closed Loop Fund (CLF)*. It is a social impact fund that invested $100 million to increase the recycling of products and packaging. It is their aim by 2025 to proof replicable recycling processes that will help unlock additional investments in recycling. [15, 16]

One of their approaches is the improvement of recycling of numbers 3 to 7 plastics. Even in major markets, no viable options for recycling these plastics exist. As a result, much of this material is being landfilled. Therefore, CLF partnered with QRS Recycling and Canusa Hershman Recycling to create a state-of-the-art plastic recovery factory in Maryland with a capacity to handle more than 50,000 t/year. This $15 million project was supported by a $2 million investment from CLF and $13 million from other private sources. [16] In QRS Recycling's Maryland facility, mixed numbers 3 to 7 plastics in bales from single-stream MRFs are transformed into a high-quality, postconsumer PP and PE flake that can be used by various plastics manufacturers. [16] One of the most significant challenges of this business model besides the sorting process is contamination. Bales of mixed plastics (numbers 3 to 7) often contain a high level of contamination. Certain materials out of these residues can have a costly impact on the process. [16]

Furthermore, the profitability of the business model is still unclear. Recycled LDPE and PP could probably be sold for $0.52 per kg (LDPE) and $0.80 per kg (PP), but the processing costs are difficult to calculate. In addition, for some feedstocks (mainly numbers 3 and 6 plastics), there is not yet any market and so value for the recycled plastics yet. Thus, these plastics would probably be landfilled, which creates negative impacts both environmentally and ecologically. [16]

However, an increasing recycling rate of plastics has to be the aim of the plastics industry. The environmental value is the highest of all optimization scenarios, mainly because plastics can be recycled several times. And even if the process might not be profitable yet, the economic potential is high.

References

[1] Kessler Consulting Inc. *MRFing Our Way to Diversion: Capturing the Commercial Waste Stream.* Tampa, FL. 2009.

[2] Pressley, P.N., Levis, J.W., Damgaard, A., Barlaz, M.A., and DeCarolis, J.F. Analysis of material recovery facilities for use in life-cycle assessment. *Waste Management.* vol. 35, pp. 307–317, 2015.

[3] Gershman; Brickner & Bratton Inc. *Materials Recovery Facility (MRF) Feasibility Report.* Fairfax, VA. 2008.

[4] Burns & McDonnell. *Mixed waste processing economic and policy study.* Austin, TX. Report prepared for the American Forest & Paper Association. 2015.

[5] Glassdoor. *Waste Management Scale Attendant Salaries.* Access Date: 2016/04/13. Available: *https://www.glassdoor.com/Salary/Waste-Management-SCALE-ATTENDANT-Salaries-E2094_D_KO17,32.htm*

[6] Salary.com. *Facility Manager Salaries.* Access Date: 2016/04/13. Available: *http://www1.salary.com/Facilities-Manager-Salary.html*

[7] Salary.com. *Heavy Equipment Operator Salaries.* Access Date: 2016/04/13. Available: *http://www1.salary.com/Heavy-Equipment-Operator-Salaries.html*

[8] Salary.com. *General Laborer Salaries.* Access Date: 2016/04/13. Available: *http://www1.salary.com/General-Laborer-Salary.html*

[9] Utracki, L. A. *Commercial polymer blends*: Springer Science & Business Media, New York, NY. 1998.

[10] Yin, S., Tuladhar, R., Shi, F., Shanks, R. A., Combe, M., and Collister, T. Mechanical reprocessing of polyolefin waste: a review. *Polymer Engineering & Science.* vol. 55, pp. 2899–2909, 2015.

[11] Salih, S. E., Hamood, A. F., and Alsabih, A. H. Comparison of the characteristics of LDPE: PP and HDPE: PP polymer blends. *Modern Applied Science.* vol. 7, p. 33, 2013.

[12] Mastalygina, E. E., Popov, A. A., Kolesnikova, N. N., and Karpova, S. G. *International Journal of Plastics Technology.* vol. 19, pp. 68–83, 2015.

[13] Axion Consulting. *Financial modelling and assessment of mixed plastic separation and reprocessing (WRAP project MDP021).* Report prepared for Waste Resources Action Programme (WRAP). 2009.

[14] Plastics News. *Current Resin Pricing.* Access Date: 2020/03/21. Available: *https://www.plasticsnews.com/resin/currentPricing/recycled-plastics*

[15] Closed Loop Fund. About the Closed Loop Fund. Access Date: 2020/05/06. Available: *https://www.closedlooppartners.com/about-us/*

[16] Closed Loop Fund. *Impact Report January 2015–September 2016.* 2016.

7 Plastic Waste around the World: Increasing Potential of Recycling

In addition to the detailed analysis of the plastics recycling market and its potential future, this book provides an outlook on waste handling and recycling in the global market. In order to understand the global effects of waste generation in general and plastics in particular, differentiating between countries by income is more useful than by geographic region. The following data was collected in the 2012 report on global solid waste management by the World Bank. [1] The numbers are only estimates because the data from some countries was missing, was from different years, and was based on slightly different assessment methodologies. Figure 7.1 shows the dependence of waste generation on income level. Low-income countries produce the least and high-income countries the most solid waste per capita. The wide ranges, such as from 0.7 to 14.0 kg/capita/day for high-income-level countries, result from disparities within the income-level groups. The waste generated is projected to grow in all geographic areas and income levels due to the increase in population and urbanization. However, the higher the income level of a country, the lower is its projected growth rate of waste generation.

Waste collection is instrumental to access the resources buried inside the waste. However, collection rates vary between 41% in low-income countries and 98% in high-income countries, mainly due to the associated cost of collection. In low-income countries, collection services account for 80 to 90% of the municipal solid waste (MSW) budget. In high-income countries, they can be as low as 10% of the MSW budget. Consumers can be required to separate their waste at the source, such as into different bins, or the unsegregated waste can be separated in sorting facilities. Developing countries use mainly single-stream systems where recyclables are collected by waste pickers during the collection process, starting prior to collection and ending at the disposal sites. In high-income countries, single-stream or multiple-stream systems, such as a combination of curbside pickup and community bins, are used, where collection is frequent and sorting facilities are highly mechanized and efficient. The total amount of recyclables and their quality depend on the degree of separation.

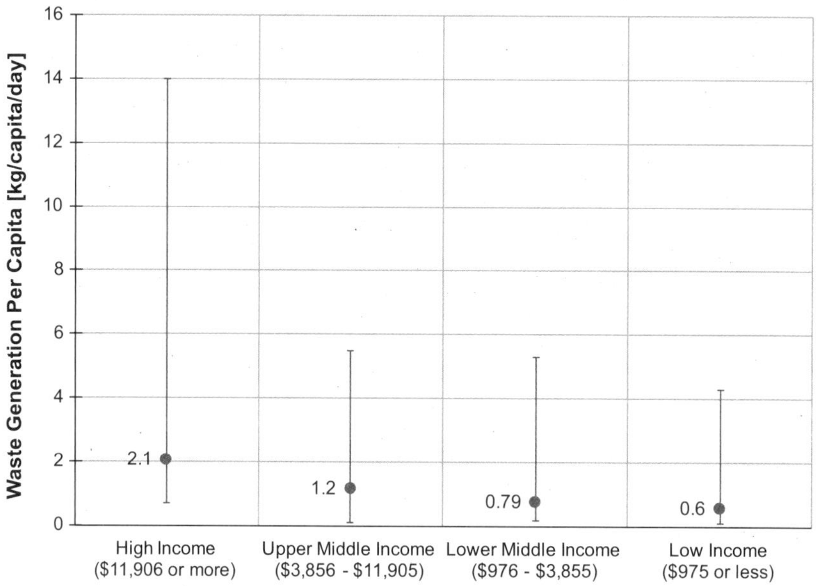

Figure 7.1 Current waste generation per capita by income level (showing the upper and lower limits and the median [dot] waste generation) [1]

The *waste composition* is important to estimate the potential of recycling valuable resources and of energy recovery. Waste composition influences the frequency of collection and disposal and is impacted by factors such as economic development, climate, energy sources, and cultural norms.

As shown in Figure 7.2, the organic fraction tends to be highest in low-income countries and lowest in high-income countries. With progressing urbanization and increase in wealth of a population, more inorganic materials (plastics, paper, and aluminum) are consumed. It is important to note that the total amount of organic waste per capita is on average still 1.5 times higher in high-income countries than in low-income countries. The same is true for all other fractions; for example, the total amount of plastic waste and paper waste is 4.9 times and 22 times higher, respectively. Geography and climate influence the waste composition. It determines the use of building materials (e.g., wood, brick, or steel), horticultural waste, and ash content. The last is related to the predominant energy source as well. Regions where energy for cooking, heating, and lighting is generated by coal and wood fires have a much higher ash content. See, for example, Figure 7.10, which shows the breakdown of waste in China for 2000, where the ash content is included in "Other".

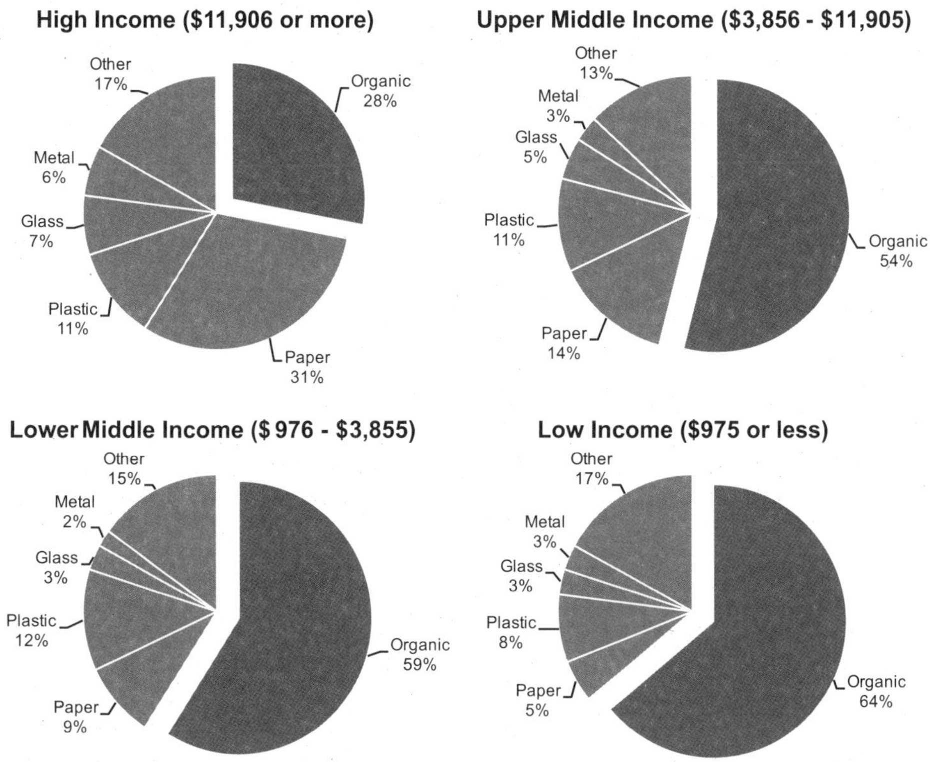

Figure 7.2 Global waste composition by income level. Organic: food scraps, yard waste, wood, and process residues; paper: paper scraps, cardboard, newspaper, magazines, bags, boxes, and wrapping paper; plastic: bottles, packaging, containers, bags, lids, and cups; glass: bottles, broken glassware, light bulbs, and colored glass; metal: cans, foils, tins, nonhazardous aerosol cans, appliances (white goods), railings, and bicycles; other: textiles, leather, rubber, multilaminates, e-waste, appliances, ash, and other inert material [1]

The waste disposal data has the highest uncertainty because most countries do not record this kind of data. MSW management practices include (a) landfilling, (b) waste-to-energy (WTE), (c) composting, (d) recycling or recovery from waste, and (e) open burning. In general, most low- and middle-income countries dispose of their MSW in open dumps. Disposal in several of the upper middle-income countries can be classified as controlled dumping, because their landfills are poorly operated. In high-income countries, landfilling and WTE are the most common methods of MSW disposal. Dumping of waste can lead to environmental burdens such as (a) contamination of groundwater and surface water by leachate, (b) soil contamination through contact with waste or leachate, (c) air pollution from uncontrolled and unfiltered burning of waste, (d) spreading of disease, (e) odor in landfills, and (f) uncontrolled release of methane by anaerobic decomposition of waste. [2]

Greenhouse gases from MSW are becoming a major concern. Postconsumer waste is estimated to contribute almost 5% (1,460 $MtCO_2E$) of the total global GHG emissions. More detailed numbers by country and composition can be found in [1] and [2].

The amount of waste, its composition, and especially the amount of plastic waste vary in different countries and regions of the world. In the further course of this chapter, plastics recycling in China and Europe as the largest and second largest plastics and product manufacturers, respectively, will serve as an example for more detailed analyses of plastics recycling. Figure 7.3 shows that the global plastics market has increased from 245 Mt in 2008 to 359 Mt in 2018, with China showing the highest market growth. It should be noted that while production volumes have increased in both Europe and the U.S., China has experienced much higher production growth, which is why the relative proportions have changed significantly. In general, the Asian market is growing steadily, accounting for over 50% of global plastic production for the first time in 2017. [3, 4]

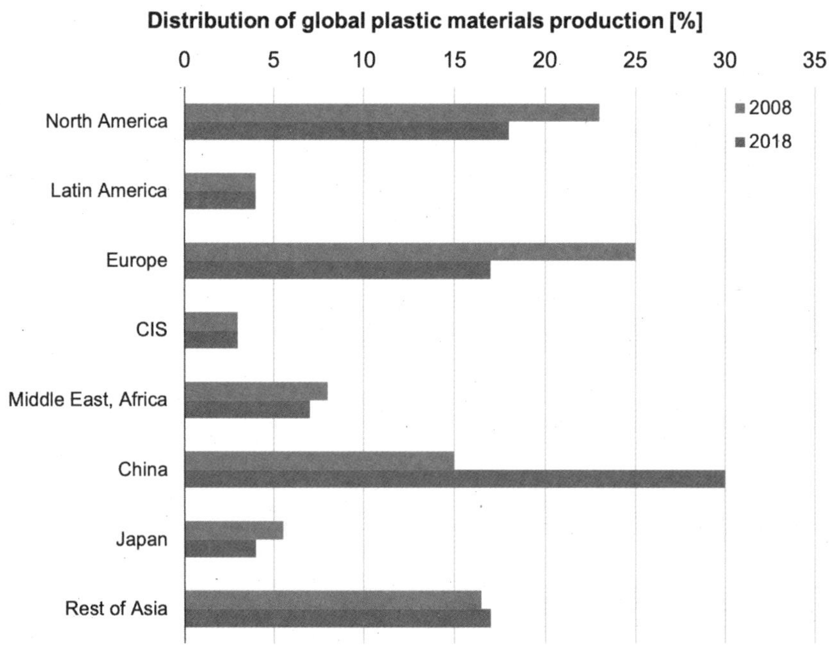

Figure 7.3 Distribution of global plastic materials production in 2008 (total: 245 Mt) and in 2018 (total: 359 Mt). [4] CIS: Commonwealth of Independent States (former Soviet Union)

7.1 Plastic Waste Handling in Europe

Europe has a very strong plastics industry, which has remained relatively consistent over the last decade. Similar to the United States, four sectors–packaging, building and construction, automotive, and electrical equipment–represent about 75% of the demand for plastics; see Figure 7.4.

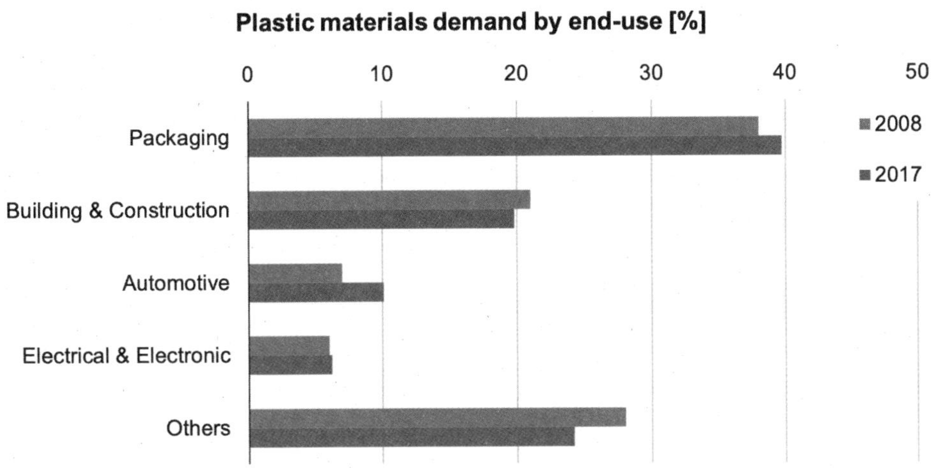

Figure 7.4 Europe's plastic materials demand by end use in 2008 (EU-27, Norway, and Switzerland) of 48.4 Mt and in 2017 (EU-28, Norway, and Switzerland) of 51.2 Mt. Others: consumer and household goods, furniture, agriculture, and medical devices [3]

Analyzing the demand for specific types of plastic, Figure 7.5 shows similar numbers as the United States. PP and PE account for almost half of the demand in comparison to all other plastics. The shares have remained almost unchanged over the last decade.

Comparing the data presented for Europe's plastic materials demand (Figure 7.4) with its generation of postconsumer waste (Figure 7.6), a difference due to the short service life of plastic packaging is observed. Although the numbers are different, the same effect was observed in the United States (Section 3.2.2.1). Other than in the United States, much of the packaging is collected from the commercial and industrial sector. From MSW, mainly PET and HDPE bottles are recovered because the contaminations (colorants and food residue) make it difficult to recycle other packaging materials.

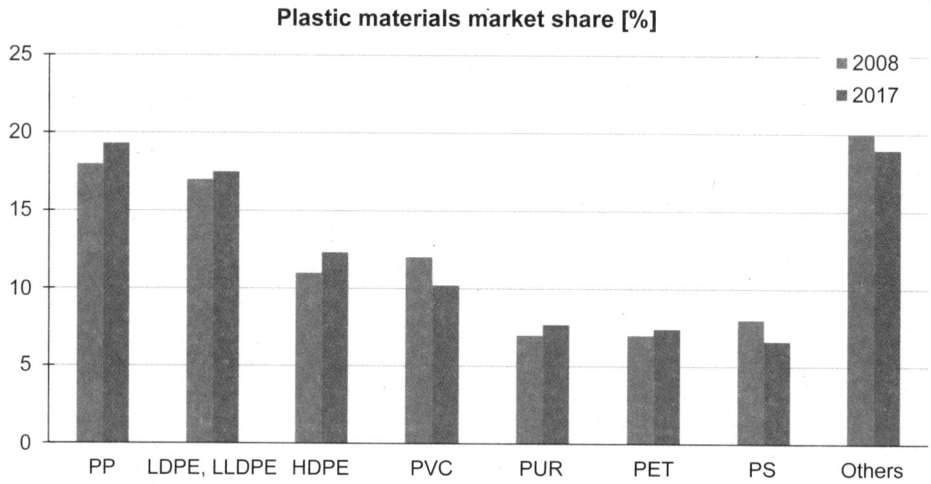

Figure 7.5 Europe's plastic materials market share in 2008 (EU-27, Norway, and Switzerland) and in 2017 (EU-28, Norway, and Switzerland) including thermoplastics, polyurethanes, and other plastics (thermosets, adhesives, coatings, and sealants) without fibers.
Others: ABS, PBT, PC, PMMA, and PTFE [3]

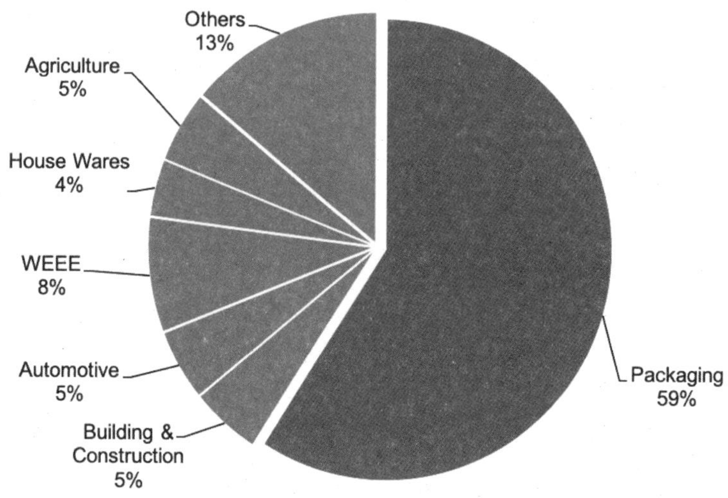

Figure 7.6 Percentages of postconsumer plastic waste in Europe (EU-27, Norway, and Switzerland) by application in 2015 [5]

In contrast, plastics used in construction have a much longer service life, between 30 to 40 years. This explains the discrepancy between the demand for plastic in this sector (19%) and the relatively low contribution (5%) to plastic waste.

Analysis of the demand for specific types of plastic in Europe (Figure 7.5) shows that PP and PE account for almost half of the demand in comparison to all other plastic materials. The shares have remained almost unchanged over the last decade.

In Europe, three waste management treatments are used: disposal in landfills, WTE, and recycling. Because of the landfill ban in Switzerland, Austria, Netherlands, Germany, Sweden, Luxembourg, Denmark, Belgium, Norway, and Finland, which started partially in 1996, landfilling has constantly declined and energy recovery and recycling have increased (Figure 7.7). In 2016, recycling overtook landfilling for the first time. In 2018, 29.1 million tons of plastic waste were collected through official schemes, from which 42.6% were burned to recover energy, 32.5% were recycled, and 24.9% were landfilled. Figure 7.7 shows the development of waste treatment in Europe between 2006 and 2018. [3]

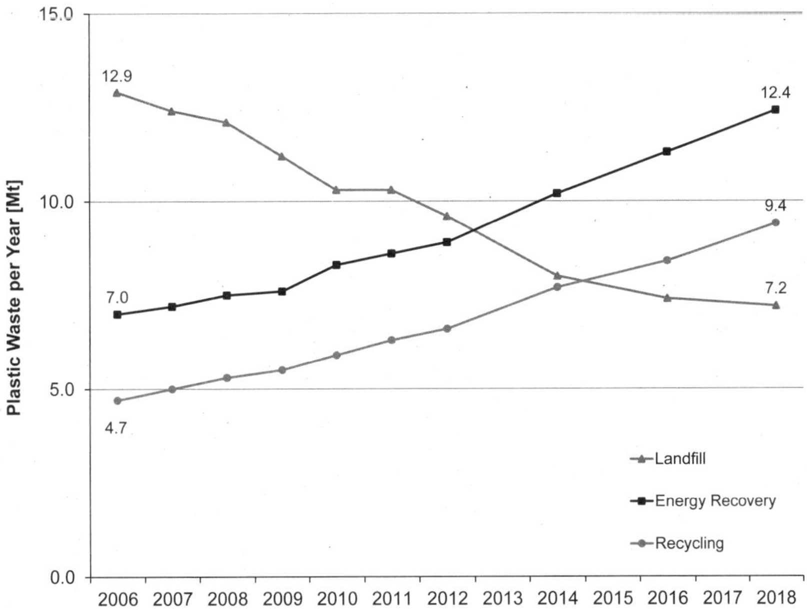

Figure 7.7 Plastic waste treatment in Europe (EU-28, Norway, and Switzerland) from 2006 to 2018 [3]

Since plastic packaging is the most significant waste stream, we will take a closer look at its treatment. To harmonize national measures concerning the management of packaging and packaging waste and to prevent or reduce its impact on the environment, the European Commission released Directive 94/62/EC on packaging and packaging waste in 1994. The Directive aims at providing a high level of environmental protection and ensuring the functioning of the internal market by avoiding obstacles to trade and distortion and restriction of competition. In 2004, the Directive was amended to provide criteria clarifying the definition of the term *packaging* and increase the targets for recovery and recycling of packaging waste. In 2005, the Directive was revised again to grant new Member States transitional periods for attaining the recovery and recycling targets. One of these targets was to increase the recycling rate of plastic packaging waste to 22.5%. [6]

The positive effect of this regulation can be seen considering the plastic packaging waste treatment in Europe between 2006 and 2018, which is shown in Figure 7.8. Despite an increase of 19.5% of total plastic packaging waste collected (14.9 Mt in 2006 to 17.8 Mt in 2018), the recycling rate increased by 92%, leading to an overall recycling rate of plastic packaging waste of 42%, fulfilling the EU directive's goal of 22.5% recycling rate. [3]

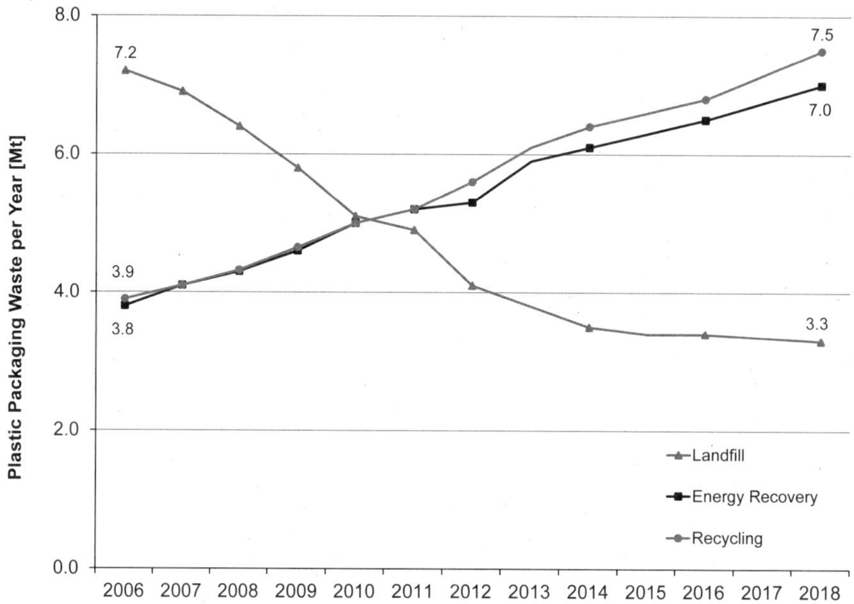

Figure 7.8 Plastic packaging waste treatment in Europe (EU-28, Norway, and Switzerland) from 2006 to 2018 [3]

However, in order to reach the target of 50% recycling rate for packaging waste, this trend has to continue within the next years. Currently, only two countries have a plastic packaging recycling rate of more than 50%: Czechia and Germany. On the contrary, plastic packaging recycling rates in Malta, Greece, France, and Finland are still lower than 25%.

In order to reach the goal of 50% recycling, different types of waste collecting systems are in place. The design of the waste flows (e.g., complete individual flow, complete separation of all recyclable materials, joint collection of materials) depends on the respective country; in Germany, even on the federal state or individual cities. In addition to door-to-door recycling flows, there are also stationary recycling containers (e.g., for paper and glass) or recycling yards (e.g., for electronic scrap). In three of the 28 capital cities of the EU, no waste is collected at all, but only collected and recycled through central points. In most countries mixed forms of recycling are implemented, as shown in Table 7.1. [7]

Table 7.1 Overview of the Collecting Systems in the 28 EU Capitals

Capital	Complete Separation (Door-to-Door)	Partial Separation/ Single-Stream (Door-to-Door)	Central Private Facility	Central Public Facility
Amsterdam	X		X	X
Athens	X	X	X	
Berlin	X	X	X	X
Bratislava			X	X
Brussels	X	X	X	X
Bucharest			X	
Budapest	X	X	X	
Copenhagen	X		X	X
Dublin	X	X	X	X
Helsinki	X		X	X
Lisbon	X		X	X
Ljubljana	X	X	X	X
London	X	X	X	X
Luxembourg	X	X	X	X
Madrid		X	X	X
Nicosia	X	X	X	X
Paris	X	X	X	X
Prague			X	X
Riga	X		X	X
Rome	X	X	X	
Sofia	X	X	X	X
Stockholm	X		X	X
Tallinn	X	X	X	X
Valletta	X	X	X	X
Vienna	X	X	X	X
Vilnius	X	X	X	X
Warsaw	X	X		
Zagreb	X	X	X	X

As the recycling system in Germany is quite successful, it is further explained in the following to highlight the differences with the U.S. system, only recycling 9% of plastic waste. Germany's dual system, chosen as Optimization Scenario I (see Section 6.1), organizes household packaging waste collection (via a yellow bag that is left curbside) and ensures presorting by the consumer and professional recycling of packaging in appropriate facilities. The disposal services (collection, sorting, and recycling of packaging) are financed via participation payments of packaging manufacturers and executed by commissioned disposal companies. Participation is mandatory and allows the manufacturers to label their products with the *green dot ("Grüner Punkt")*. Since manufacturers are required to pay for their packaging products by weight, material consumption has decreased since the system was introduced due to use of thinner glass, paper, plastic bottles, and films. This further reduces the amount of waste that needs recycling.

As already mentioned in Chapter 2, for both glass and plastic bottles a *deposit and return system* exists. The deposit is paid by the consumer upon purchase of a product. It ranges from €0.15 for refillable bottles to €0.25 for disposable bottles. The empty bottle can be returned to stations similar to vending machines that automatically scan and accept or reject the bottle and reimburse the deposit for all refillable bottles. Due to their light weight and the reduction of use of virgin plastics, the refillable PET bottles are slightly more ecologically friendly than refillable glass bottles and significantly more so than disposable bottles.

All these improvements are possible due to mechanized collection, efficient recycling, waste treatment facilities, and continued education of the consumer over the last three decades. Last but not least, the economic strength of Germany and other European countries enables the investment in such projects and initiatives.

7.2 Plastic Waste Handling in China

As a consequence of China's fast economic growth, in 2013 the country bypassed all other countries and regions and became the world's largest producer of plastic products. Its domestic consumption of plastics has skyrocketed as well. [8] China is also the main market for postconsumer plastic waste. The demand for postconsumer plastic waste is spread across seven main sectors of the economy, with the top three, packaging, construction, and appliances, accounting for almost 80% of the total demand, as shown in Figure 7.9. However, there are different estimates of the balance of end uses of these materials. For example, the proportion of postconsumer plastic waste used in textiles could be significantly higher than that shown in the graph and packaging accordingly lower. [9]

The packaging sector accounts for over 40% of the demand for postconsumer plastic waste in China, mainly for use in the production of bottles and bags. Polyethylene (PE) is typically used to make bottles for shampoo, detergent, and industrial oil. Recycled PET is almost exclusively used in fiber applications, but its use in packaging is now growing as well. Polypropylene (PP) is used to make cement and fertilizer bags. [9]

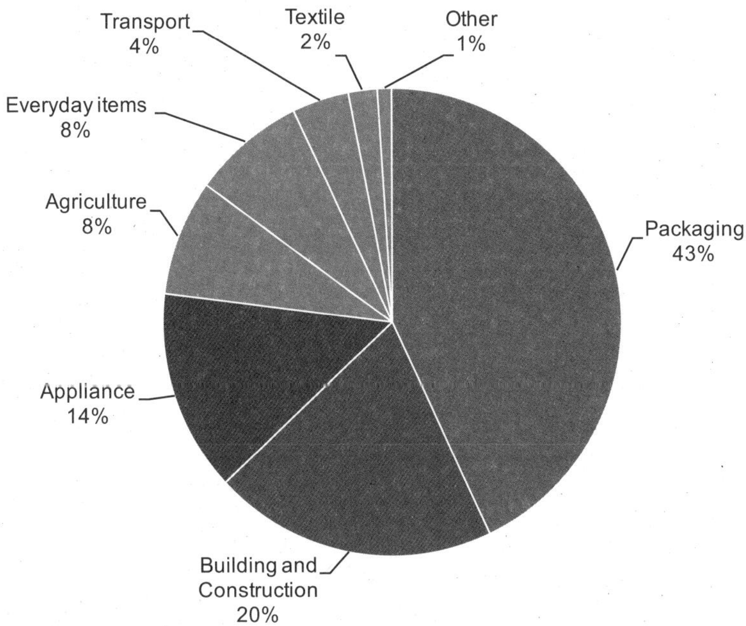

Figure 7.9 Postconsumer plastic waste demand by sector in China in 2011. Courtesy of CBI (China) Co., Ltd., Shanghai, China [6]

Construction has been a fast-growing sector in China, accounting for about 20% of the demand, mostly to make pipes and boards. The main postconsumer plastic waste used is recycled polyvinyl chloride (PVC), but small amounts of PE and PP are also used. Extruded polystyrene (XPS) board is used as insulating material for walls and roofs. Waterproof film to protect construction sites from the weather is made from recycled PE. There are some newer applications, such as wood-plastic composites in which postconsumer plastic waste (again recycled PE and recycled PP) are compounded with wood fibers. [9]

In-country recycling in China is, however, much more difficult than in Europe and the U.S. In many places the infrastructure is not yet as well developed as in the other regions mentioned. The market for recycled plastics is still dominated by a large number of small manufacturers and recyclers who work with low-tech equipment and without any environmental protection practices. They also produce low-quality recycled products. The reason for this is the undeveloped plastic recycling process: Plastic waste is collected by individual bottle collectors, people who offer small sums of money to households for their recyclable products. They sell them to recycling companies, who in turn sell them to the reprocessing companies. However, the quality of the recycled collection materials is not uniform. As a result, the reprocessors can only produce inferior products in large quantities, such as plastic bags and disposable tableware. Some of these cheap products are sent back to Europe and North America. However, the majority are consumed in China. [10]

The quantity and quality of plastic waste is also strongly influenced by the energy source. Figure 7.10 shows the difference in the share of plastic waste in total municipal waste, depending on whether coal or gas is used for heating. For example, the share of plastic waste in coal-fired households was 4% of municipal waste, which represented about 2 Mt of plastic waste. In gas-fired households the share was much higher at 13%, representing about 13 Mt of plastic waste. The "other" waste, which is significantly higher in coal-fired households, often contains harmful waste, which in turn contaminates plastics and reduces their quality. [11]

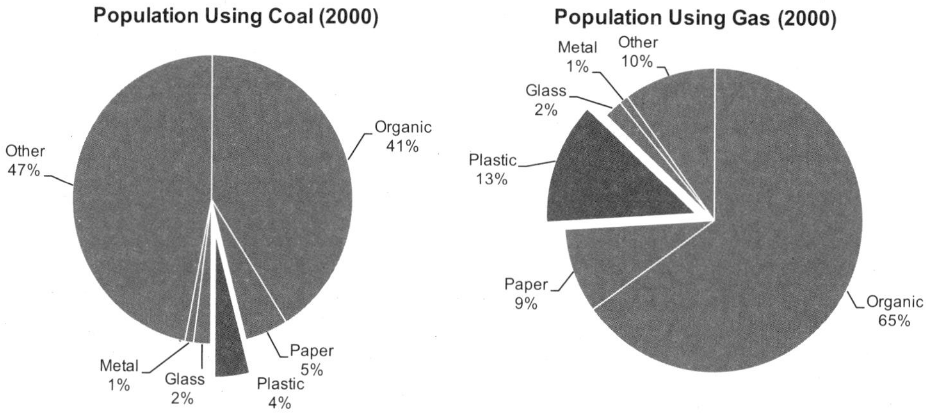

Figure 7.10 Municipal waste composition in China in 2000 [7]

Due to the low quality of domestic postconsumer plastic waste sources and the high demand for plastics, China has become the world's largest importer of plastics in the first 15 years of the 21st century. Due to the high quality of *imported postconsumer plastic* waste sources, many Chinese recyclers have not considered sourcing plastic waste domestically. For example, the import quota in China–despite a domestic plastic waste volume of about 60 Mt–was only marginally lower than the domestic volume of recycled plastic waste, especially until 2009, as Figure 7.11 shows. Improved technologies and growing environmental awareness led to a 50% increase in the amount of recycled plastics between 2009 and 2011, which is disproportionately high compared to the increase in domestic waste volumes. Nevertheless, an increase in the import quota was still recorded. [9, 12, 13, 14, 15]

However, imports of plastic waste are not only of great importance for Chinese producers, but also for the countries and regions exporting to China. For example, by 2017 China had processed approximately 56% of the world's recycled plastic waste and purchased large amounts of recycled waste, particularly from the industrial regions of Europe and the U.S. [16]

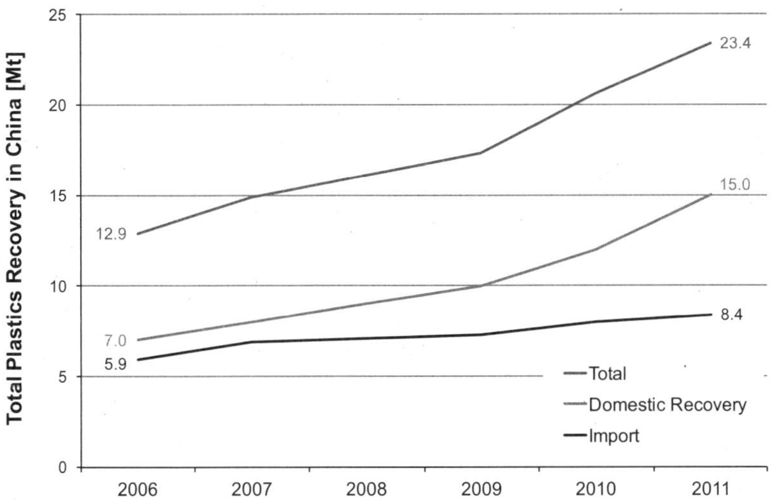

Figure 7.11 Total postconsumer plastic waste recovery in China from 2006 to 2011 [15]

Since 2013, however, the import quota of postconsumer plastic waste has been declining–despite the continuing increase in demand. This effect results from an initiative of *Operation Green Fence* in 2010, aiming to reduce the amount of poor-quality plastic waste imported into China but only importing plastics that have already been sorted by type. As a result, a total of 2 million tons of postconsumer plastic waste was imported in the period from January to April 2016, a decrease of 11% compared to the same period in 2015. [10, 12, 17]

In 2017, this law was tightened considerably. A stricter, permanent *ban on non-industrial plastics imports* was introduced. The decision shook the global recycling system and heralded a new era in waste management. As a result, many of the world's industrialized nations, including the U.S., Canada, Europe, Japan, and South Korea, are being forced to look for new ways to use their recycled waste.

In the first year of the ban, the dimension of this decision already became apparent: As shown in Figure 4.7, the plastics exports of the U.S. from January to June 2017 and January to June 2018 have decreased by about one third, from 950,000 tons to 660,000 tons. This is nearly the exact difference as the decrease of U.S. exports to China, which recorded a drop from 380,000 tons (January to June 2017) to 35,000 tons (January to June 2018). A similar trend is evident in Europe. Exports of the United Kingdom fell from around 90,000 tons (January to April 2017) to around 1,500 tons (January to April 2018). Small amounts of the plastic waste were exported to other, primarily Asian, countries (e.g., Malaysia, Taiwan, Vietnam, Indonesia). In these countries, however, the waste is often incinerated with open fires and the waste problem is merely shifted. [13, 16, 18, 19]

Thus, new plastic waste handling measures need to be developed quickly, as the cumulative displacement of global plastic waste is expected to reach an estimated 110 million tons by 2030, depending on the severity with which China imposes the import ban. Figure 7.12 illustrates this for three scenarios, namely the assumption of a 50-, 75-, and 100-percent import ban. [13]

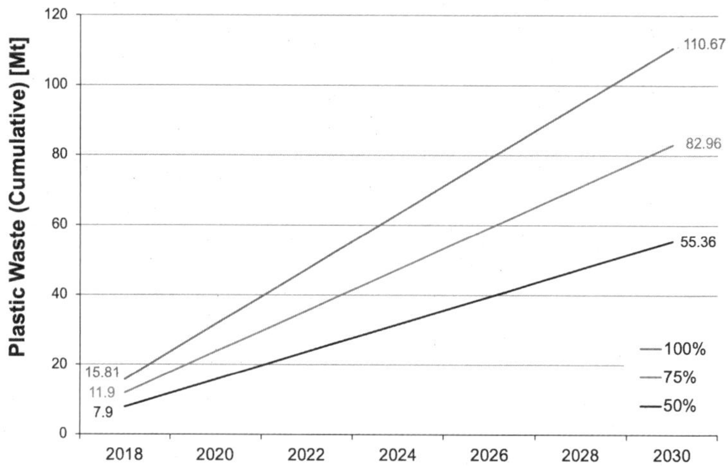

Figure 7.12 Cumulative amount of plastic waste resulting from the Chinese import ban, presented for three possible import ban scenarios (50%, 75%, and 100%) [15]

Despite the import ban on plastic waste, China is expected to be able to completely cover its demand for postconsumer waste through domestic recycling by 2030. The predicted municipal waste composition in China in 2030 is shown in Figure 7.13. With a predicted waste volume of 485 million tons in 2030 including a total of 14% plastic waste, 68 tons of plastic waste will be produced each year in China. With the increasing efficiency of the collection process, this has enormous potential to meet China's domestic demand for plastic waste.

Thus, China is not only world leader in producing plastic products, but also in producing plastic waste. Due to the currently low recycling rates of domestic plastic waste, China is also responsible for an estimated 28% of the plastic waste in the world's waters. With a projected increase in the Chinese population of 300 million inhabitants by 2025 and the associated increase in plastic waste to 70 million tons by 2030, proper waste management will play a key role in China–for China itself and the world as a whole.

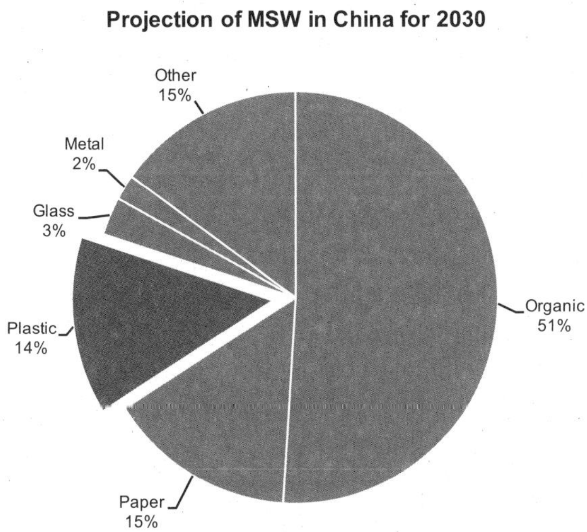

Figure 7.13 The estimated municipal solid waste (MSW) composition in China in 2030 [11]

It is important to maintain certain quality standards in recycling. Currently, many Chinese recycling companies add large quantities of additives, such as cheap fluorosurfactants, to plastics during the processing stage. Although this improves the material properties, it is very damaging to the environment, should the material be recycled again. After all, the materials supplemented by additives are almost impossible to recycle, which means that valuable resources are lost. [8, 20]

However, China is implementing recycling projects. Shanghai was the first city to launch a major *recycling initiative* in June 2019, which will have an impact on waste separation for the end consumer. Whereas previously all MSW was disposed of in one bin, Shanghai has now added four bins: one bin for dry residual waste, one bin for wet residual waste, one bin for hazardous materials, and one bin for recycled material. The bins are emptied every day. In addition, according to media reports, around 30,000 volunteers are on their way to explain waste separation to Shanghai residents. The first positive effects were quickly noticed, especially with regard to the reduction of waste on the streets. [21]

Despite improved recycling processes and waste import bans, the waste problem in China is still very serious. The large number of landfills must be reduced as quickly as possible. For this reason, China has invested heavily in WTE facilities in recent years. This solves the problem of landfill space limitations. In addition, the incineration process serves as an important source of energy. China now has over 300 WTE plants, which is the largest number in the world. This large number is the result of a strong expansion in the last five years, with the number of plants increasing by 26% annually. In comparison, the OECD countries have reported an average increase of 4% over the same period. China is focusing on large, innovative plants, as shown by the *Shenzhen East Waste-to-Energy Plant.* This plant is scheduled to go operational in 2020 and will be the largest in the world, processing 5,000 tons of waste per day. In addition, the 66,000 m^2 roof is to be covered with up to 44,000 m^2 of photovoltaic modules and will enable the plant not only to provide cleaner treatment of municipal waste, but also to contribute to the provision of renewable energy for the city. [22]

However, in incineration—as in recycling—attention must also be paid to the cleanliness and efficiency of the process. Unsorted Chinese municipal waste with a high proportion of moist organic material reduces the efficiency of incineration, as explained in detail in the economic analysis in Section 4.3. Significantly more fuel is required for the incineration of moist waste, which increases costs and reduces or even nullifies the profits from energy generation. Another problem is the poor regulation of waste incineration plants in China. Although highly innovative plants already exist, as shown by the example of the Shenzhen East Waste-to-Energy Plant, there are still very poorly regulated plants in operation. The resulting toxic air pollution in turn leads to environmental and health problems. An alternative or supplement to incineration is the use of anaerobic bacteria to decompose organic waste and even to separate the methane as a fuel source, which has recently attracted some interest. [10, 23, 24]

7.3 Plastic Waste in the Future

Due to the global growth of plastic materials in many industry sectors, the generation of plastic waste has increased. From 1950 to 2012, the growth of the industry was 8.7% on average per year, which equals a total increase from 1.7 Mt in 1950 to the 359 Mt in 2018. [3, 25] In parallel, the percentage of plastic waste in MSW grew as well, which demonstrates the rising significance of plastics in our daily life. Today, an average person living in Western Europe or North America consumes 100 kg of plastic each year, mostly in the form of packaging. Asia uses just 20 kg per person, but this number is predicted to grow rapidly as economies in the region expand. [19] Our standard of living directly correlates to the amount of plastic in our MSW. Waste generation rates tend to be much higher in urban centers. Today, more than 50% of the world's population lives in cities, and the rate of urbanization is increasing quickly. By 2050, as many people will live in cities as the population of the whole world in 2000. This will add challenges to waste disposal. [1, 25]

As shown in Chapter 4, recovering plastic from the waste stream for recycling or for incineration with energy recovery has the potential to minimize these problems. As of today, only about 10% and 30% of the plastic in MSW is recycled in the United States and Europe, respectively. In addition, much of the recovered plastic is exported to countries with lower environmental regulations, which was noted for the example of Asia. Furthermore, WTE facilities require air emissions controls and produce hazardous ash. [25]

Chapter 5 explains why recycling is the necessary method of handling our plastic waste, even if the process is currently not profitable. However, the two optimization scenarios in Chapter 6 showed that plastic recycling has the potential to be profitable if the process is changed or extended. In the future, this potential will increase, both for the United States and worldwide.

Based on the growth predictions, citizens and corporations will likely need to assume more responsibility for waste generation and disposal, specifically, in terms of product design and waste separation. Also likely to emerge will be a greater emphasis on "urban mining". [1]

Figure 7.14 shows an even broader view of the economic potential of plastics recycling, identified by Project MainStream, a multi-industry, global initiative launched in 2014 by the World Economic Forum and the Ellen MacArthur Foundation, with McKinsey & Company as knowledge partner. Currently, only 5% of the material value of plastic packaging is captured worldwide after one use, corresponding to $4 billion to $6 billion per year. The actual value yield of recycling is only 36%. By blending plastics, such as LDPE and PP during recycling, and thereby improving the properties of the recycled material, this yield can be increased significantly. Furthermore, the current worldwide plastic recycling rate is only 14%. Levers for an enhanced recycling rate are secondary markets, a global plastic protocol, technological innovations, and enabling policies. By increasing both the value yield and the recycling rate, the captured material value could be between $80 billion and $120 billion per year. [25]

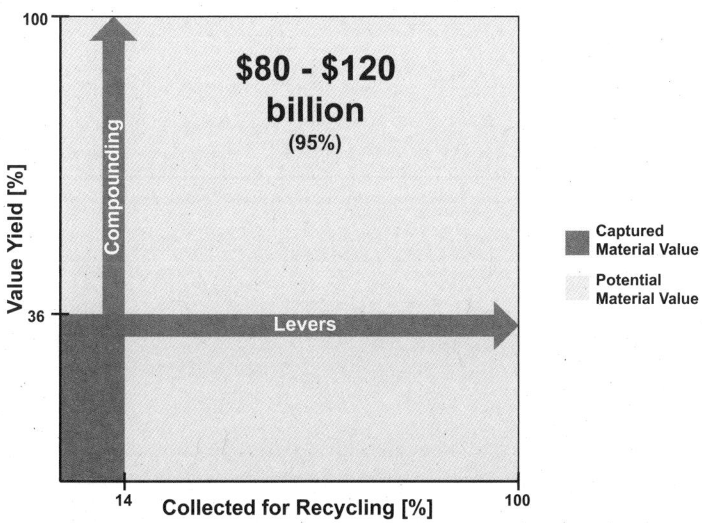

Figure 7.14 Economic potential of recycling plastic waste [16]

Besides the approach of compounding, the ongoing *digital transformation* promises to support recycling activities. By gathering and using data over the product lifecycle–starting from production, over usage, collection, recycling, and re-usage–the product circle could be closed, and the recycling rate increased. Until now, it has not been possible to close this circle and overcome the information deficit that also hampers the plastic recycling. However, some solutions are already in place, which in sum could support the recycling process. These are explained in the following. [26, 27]

Cyber physical systems allow products to carry information at least through the entire production process. If the exact information regarding the material composition of a product is recorded and somehow traceable for a product, this would simplify recycling. It would also be the basis for *a data management infrastructure* to track individual consumer products through the global supply chain–from design to creation, to use, to disposal. [26, 27] *Sensoring* enables the real-time collection and generation of data. Up-to-date, detailed information about the precise location of waste production, its exact material composition, etc. could be recorded and passed on to other companies, which then use it to plan their production processes. [26] Furthermore, if data were available for the consumer as well, a closed-loop process for recycling that both *encourages and rewards participation* by consumers and supports the re-use of materials in a circular economy could be implemented. [27] These are only a few solutions for how digitalization could increase the recycling–but they are also still in development.

However, this potential analysis of both compounding and digitalization reflects the results of Chapter 6 of this book: recycling is not only the best waste handling method for our environment, but also its economic potential value is high. It is the duty of our society to exploit the maximum potential of the plastic recycling process.

References

[1] Hoornweg, D., and Bhada-Tata, P. What a waste: a global review of solid waste management. *Urban Development Series Knowledge Papers.* vol. 15, pp. 1–98, 2012.

[2] Karak, T., Bhagat, R., and Bhattacharyya, P. Municipal solid waste generation, composition, and management: the world scenario. *Critical Reviews in Environmental Science and Technology.* vol. 42, pp. 1509–1630, 2012.

[3] PlasticsEurope. *Plastics–the Facts 2019.* Brussels, Belgium. 2020.

[4] Statista. *Verteilung der weltweiten Kunststoffproduktion nach Regionen 2018.* Access Date: 2020/03/21. Available: *https://de.statista.com/statistik/daten/studie/244172/umfrage/verteilung-der-weltweiten-kunststoffproduktion-nach-regionen/*

[5] European Commission. *A European Strategy for Plastics in a Circular Economy.* Access Date: 2020/03/21. Available: *https://eur-lex.europa.eu/legal-content/EN/TXT/HTML/?uri=CELEX:52018DC0028&from=EN*

[6] European Commission. Packaging and Packaging Waste. Access Date: 2020/03/21. Available: *https://ec.europa.eu/environment/waste/packaging/index_en.htm*

[7] Bipro. *Assessment of separate collection schemes in the 28 capitals of the EU.* Final Report. Copenhagen, 2015.

[8] Collective Responsibility. *Plastic Waste in China.* Shanghai, China. Access Date: 2016/12/05. Available: *http://www.coresponsibility.com/plastic-waste-in-china/*

[9] Waste and Resource Action Programme (WRAP). *The Chinese markets for recovered paper and plastics–an update.* Market situation report. pp. 1–15, Spring 2011.

[10] Koty, A.C. *Trash or Treasure?–Prospects for China's Recycling Industry.* China Briefing, Dezan Shira & Associates. pp. 1–7, 2016.

[11] The World Bank. *Waste Management in China: Issues and Recommendations: May 2005.* Urban Development Working Papers, East Asia Infrastructure Department, The World Bank, Washington, D.C. Working Paper No. 9. 2005.

[12] Waste and Resource Action Programme (WRAP). *Plastics Market Situation Report. Spring 2016.* pp. 1–29.

[13] Brooks, A.L., Wang, S., and Jambeck, J.R. *The Chinese import ban and its impact on global plastic waste trade.* 2018.

[14] Waste and Resource Action Programme (WRAP). *Market Snapshot – June 2016.* 2016.

[15] Velis, C. *Global recycling markets: plastic waste: A story for one player–China.* International Solid Waste Association. Vienna, Austria. 2014.

[16] Deutsche Welle. *After China's import ban, where to with the world's waste?* Access Date: 2020/03/21. Available: *https://www.dw.com/en/after-chinas-import-ban-where-to-with-the-worlds-waste/a-48213871*

[17] The World Bank. *Decision Makers' Guide to Municipal Solid Waste Incineration.* The World Bank, Washington, D.C. 1999.

[18] Clarke, J.S., and Howard, E. *US Plastic Waste Exports to Developing Countries, Causing Environmental Problems at Home and Abroad.* Access Date: 2020/03/21. Available: *https://unearthed.greenpeace.org/2018/10/05/plastic-waste-china-ban-united-states-america/*

[19] Financial Times. *Plastic waste export tide turns to south-east Asia after China ban.* Access Date: 2020/03/21. Available: *https://www.ft.com/content/94ee72d0-6f26-11e8-852d-d8b934ff5ffa*

[20] Jambeck, J.R., Geyer, R., Wilcox, C., Siegler, T.R., Perryman, M., Andrady, A., Narayan, R., and Law, K.L. Marine pollution: plastic waste inputs from land into the ocean. *Science.* vol. 347, pp. 768–771, 2015.

[21] Wurzel, S. *Shanghai entdeckt das Recycling - Umdenken bei der Entsorgung.* Access Date: 2020/03/21. Available: *https://www.tagesschau.de/ausland/china-recycling-101~amp.html*

[22] Wood, J. *This Chinese megacity is building a giant waste-to-energy plant.* World Economic Forum. Access Date: 2020/03/21. Available: *https://www.weforum.org/agenda/2019/07/one-of-china-s-biggest-megacities-is-building-a-giant-waste-to-energy-plant*

[23] Look, M. *Trash Planet: China.* Earth 911. pp. 1–11, 2009.

[24] Engberg, G. Review of Mitchell Young: *Garbage and Recycling.* In: *Booklist.* vol. 104, p. 35, 2007.

[25] Project MainStream. *The New Plastics Economy: Rethinking the Future of Plastics.* Ellen MacArthur Foundation, The World Economic Forum, and McKinsey & Company. 2016.

[26] Wilts, H. *The digital circular economy: can the digital transformation pave the way for resource-efficient materials cycles?* Sustainability Impulses from Wuppertal. 2017.

[27] Murphy, N. *Here's how digitalization can boost recycling rates.* World Economic Forum. Access Date: 2020/03/21. Available: *https://www.weforum.org/agenda/2019/01/here-s-how-digitization-can-boost-recycling-rates/*

8 Appendix

This chapter shows the detailed numbers and calculations of the economic analysis in Chapter 4 and the optimization scenarios in Chapter 6.

Since these calculations are based and depend on several assumptions (e.g., efficiencies, size of factories, waste capacity, exchange rate, salaries), the tables enable the analysis of different scenarios for landfilling, incineration with energy recovery, and recycling of plastic waste.

> Yet more detailed spreadsheets used in these analyses can be downloaded from *https://plus.hanser-fachbuch.de/en,* by entering the code provided on the first page of this book. These can be adapted to allow you to obtain results for your own ideas and scenarios.

8.1 Economic Analysis of Landfilling

Table 8.1 Economic Analysis of Landfill: Construction Costs (imperial)

Buildings	**Area [sq ft]**	**Costs [$/sq ft]**			**Total Costs [$]**		
		Min.	**Max.**	**Average**	**Min.**	**Max.**	**Average**
Maintenance Buildings	10,000	50.00	70.00	60.00	500,000.00	700,000.00	600,000.00
Office Buildings	3,000	60.00	100.00	80.00	180,000.00	300,000.00	240,000.00
Shacks and Tool Sheds	1,000	10.00	20.00	15.00	10,000.00	20,000.00	15,000.00
IT	**Pieces**	**Costs per Unit [$/piece]**			**Total Costs [$]**		
		Min.	**Max.**	**Average**	**Min.**	**Max.**	**Average**
Associated Computer Systems	1	50,000.00	75,000.00	62,500.00	50,000.00	75,000.00	62,500.00
Modular Truck Scales	1	50,000.00	75,000.00	62,500.00	50,000.00	75,000.00	62,500.00
Roads	**Area [sq ft]**	**Costs [$/sq ft]**			**Total Costs [$]**		
		Min.	**Max.**	**Average**	**Min.**	**Max.**	**Average**
Perimeter Access Roads (Gravel)	120,000	1.00	2.00	1.50	120,000.00	240,000.00	180,000.00
Perimeter Access Roads (Asphalt)	0	6.00	9.00	7.50	0.00	0.00	0.00
Security Barrier	**Length [ft]**	**Costs [$/.ft.]**			**Total Costs [$]**		
		Min.	**Max.**	**Average**	**Min.**	**Max.**	**Average**
Fence	6,000	10.00	20.00	15.00	60,000.00	120,000.00	90,000.00
Gates	5	1,000.00	2,000.00	1,500.00	5,000.00	10,000.00	7,500.00
Signages (200 ft Intervals)	30 pc.	10.00	20.00	15.00	300.00	600.00	450.00
Washing	**Pieces**	**Costs per Unit [$/piece]**			**Total Costs [$]**		
		Min.	**Max.**	**Average**	**Min.**	**Max.**	**Average**
Wheel Washing Facilities	1	200,000.00	250,000.00	225,000.00	200,000.00	250,000.00	225,000.00
Liner Construction	**Area [acre]**	**Costs [$/acre]**			**Total Costs [$]**		
		Min.	**Max.**	**Average**	**Min.**	**Max.**	**Average**
Clear and Grub	33.50	1,000.00	3,000.00	2,000.00	33,500.00	100,500.00	67,000.00
Site Survey	33.50	5,000.00	8,000.00	6,500.00	167,500.00	268,000.00	217,750.00
Excavation	33.50	100,000.00	330,000.00	215,000.00	3,350,000.00	11,055,000.00	7,202,500.00
Perimeter Berm	33.50	10,000.00	16,000.00	13,000.00	335,000.00	536,000.00	435,500.00
Clay Liner	33.50	32,000.00	162,000.00	97,000.00	1,072,000.00	5,427,000.00	3,249,500.00
Geomembrane	33.50	24,000.00	35,000.00	29,500.00	804,000.00	1,172,500.00	988,250.00
Geocomposite	33.50	33,000.00	44,000.00	38,500.00	1,105,500.00	1,474,000.00	1,289,750.00
Granular Soil	33.50	48,000.00	64,000.00	56,000.00	1,608,000.00	2,144,000.00	1,876,000.00
Leachate System	33.50	8,000.00	102,000.00	55,000.00	268,000.00	3,417,000.00	1,842,500.00
QA/QC	33.50	75,000.00	100,000.00	87,500.00	2,512,500.00	3,350,000.00	2,931,250.00
Total Costs					**12,431,300.00**	**30,734,600.00**	**21,582,950.00**

Table 8.2 Economic Analysis of Landfill: Construction Costs (metric)

Buildings	**Area [sq m]**	**Costs [$/sq m]**			**Total Costs [$]**		
		Min.	**Max.**	**Average**	**Min.**	**Max.**	**Average**
Maintenance Buildings	929	538.18	753.46	645.82	500,000.00	700,000.00	600,000.00
Office Buildings	279	645.82	1,076.36	861.09	180,000.00	300,000.00	240,000.00
Shacks and Tool Sheds	93	107.64	215.27	161.45	10,000.00	20,000.00	15,000.00
IT	**Pieces**	**Costs per Unit [$/piece]**			**Total Costs [$]**		
		Min.	**Max.**	**Average**	**Min.**	**Max.**	**Average**
Associated Computer Systems	1	50,000.00	75,000.00	62,500.00	50,000.00	75,000.00	62,500.00
Modular Truck Scales	1	50,000.00	75,000.00	62,500.00	50,000.00	75,000.00	62,500.00
Roads	**Area [sq m]**	**Costs [$/sq m]**			**Total Costs [$]**		
		Min.	**Max.**	**Average**	**Min.**	**Max.**	**Average**
Perimeter Access Roads (Gravel)	11,149	10.76	21.53	16.15	120,000.00	240,000.00	180,000.00
Perimeter Access Roads (Asphalt)	0	64.58	96.87	80.73	0.00	0.00	0.00
Security Barrier	**Length [m]**	**Costs [$/m]**			**Total Costs [$]**		
		Min.	**Max.**	**Average**	**Min.**	**Max.**	**Average**
Fence	1,829	32.81	65.62	49.21	60,000.00	120,000.00	90,000.00
Gates	2	3,280.80	6,561.60	4,921.20	5,000.00	10,000.00	7,500.00
Signages (200 ft Intervals)	30 pc.	10.00	20.00	15.00	300.00	600.00	450.00
Washing	**Pieces**	**Costs per Unit [$/piece]**			**Total Costs [$]**		
		Min.	**Max.**	**Average**	**Min.**	**Max.**	**Average**
Wheel Washing Facilities	1	200,000.00	250,000.00	225,000.00	200,000.00	250,000.00	225,000.00
Liner Construction	**Area [sq m]**	**Costs [$/sq m]**			**Total Costs [$]**		
		Min.	**Max.**	**Average**	**Min.**	**Max.**	**Average**
Clear and Grub	13.56	2,471.05	7,413.16	4,942.11	33,500.00	100,500.00	67,000.00
Site Survey	13.56	12,355.27	19,768.43	16,061.85	167,500.00	268,000.00	217,750.00
Excavation	13.56	247,105.41	815,447.84	531,276.63	3,350,000.00	11,055,000.00	7,202,500.00
Perimeter Berm	13.56	24,710.54	39,536.87	32,123.70	335,000.00	536,000.00	435,500.00
Clay Liner	13.56	79,073.73	400,310.76	239,692.25	1,072,000.00	5,427,000.00	3,249,500.00
Geomembrane	13.56	59,305.30	86,486.89	72,896.10	804,000.00	1,172,500.00	988,250.00
Geocomposite	13.56	81,544.78	108,726.38	95,135.58	1,105,500.00	1,474,000.00	1,289,750.00
Granular Soil	13.56	118,610.60	158,147.46	138,379.03	1,608,000.00	2,144,000.00	1,876,000.00
Leachate System	13.56	19,768.43	252,047.52	135,907.97	268,000.00	3,417,000.00	1,842,500.00
QA/QC	13.56	185,329.06	247,105.41	216,217.23	2,512,500.00	3,350,000.00	2,931,250.00
Total Costs					**12,431,300.00**	**30,734,600.00**	**21,582,950.00**

Table 8.3 Economic Analysis of Landfill: Operations Costs

Personnel	**Units**	**Unit Costs [$/unit]**	**Total Costs [$]**
Facility Manager	2	90,000.00	180,000.00
Equipment Operators	4	60,000.00	240,000.00
Scale House Attendant	2	40,000.00	80,000.00
General Laborers	4	30,000.00	120,000.00
Equipment Operating Costs	**Usage [h]**	**Cost per Hour [$/h]**	**Total Costs [$]**
Bulldozer	3,180	55.00	174,900.00
Compactor	3,180	50.00	159,000.00
Front-End-Loader	3,180	50.00	159,000.00
Grader	3,180	30.00	95,400.00
Site Repairs and Maintenance	**Units**	**Unit Costs [$/unit]**	**Total Costs [$]**
Perimeter Access Roads (Gravel)	1	160,000.00	160,000.00
Leachate Collection and Treatment	**Gallons**	**Costs per Gallon [$/gallon]**	**Total Costs [$]**
Leachate Collection and Treatment Costs	1,500,000.00	0.02	30,000.00
Others	**Units**	**Unit Costs [$/unit]**	**Total Costs [$]**
Environmental Sampling and Monitoring	1	60,000.00	60,000.00
Engineering Services (Consulting, in-house staff)	1	120,000.00	120,000.00
Operation Costs per Year [$]			**1,578,300.00**
Total Operation Costs (11 years) [$]			**17,361,300.00**

Table 8.4 Economic Analysis of Landfill: Closure Costs (imperial)

Task	Area [acre]	Costs [$/acre]			Total Costs [$]		
		Min.	Max.	Average	Min.	Max.	Average
Final Grade Survey	34.00	3,000.00	6,000.00	4,500.00	102,000.00	204,000.00	153,000.00
Gas Management Layer	34.00	24,000.00	32,000.00	28,000.00	816,000.00	1,088,000.00	952,000.00
Compacted Clay Cap	34.00	26,000.00	51,000.00	38,500.00	884,000.00	1,734,000.00	1,309,000.00
Geomembrane Cap	34.00	18,000.00	23,000.00	20,500.00	612,000.00	782,000.00	697,000.00
Geocomposite	34.00	33,000.00	44,000.00	38,500.00	1,122,000.00	1,496,000.00	1,309,000.00
Cover and Vegetative Soil	34.00	13,000.00	26,000.00	19,500.00	442,000.00	884,000.00	663,000.00
Seed, Mulch, and Fertilizer	34.00	1,000.00	2,000.00	1,500.00	34,000.00	68,000.00	51,000.00
Gas Management System	34.00	29,000.00	35,000.00	32,000.00	986,000.00	1,190,000.00	1,088,000.00
Run-Off Control System	34.00	5,000.00	7,000.00	6,000.00	170,000.00	238,000.00	204,000.00
QA/QC	34.00	75,000.00	100,000.00	87,500.00	2,550,000.00	3,400,000.00	2,975,000.00
Total		**227,000.00**	**326,000.00**	**276,500.00**	**7,718,000.00**	**11,084,000.00**	**9,401,000.00**

Table 8.5 Economic Analysis of Landfill: Closure Costs (metric)

Task	Area [sq m]	Costs [$/sq m]			Total Costs [$]		
		Min.	Max.	Average	Min.	Max.	Average
Final Grade Survey	13.76	7,413.16	14,826.32	11,119.74	102,000.00	204,000.00	153,000.00
Gas Management Layer	13.76	59,305.30	79,073.73	69,189.51	816,000.00	1,088,000.00	952,000.00
Compacted Clay Cap	13.76	64,247.41	126,023.76	95,135.58	884,000.00	1,734,000.00	1,309,000.00
Geomembrane Cap	13.76	44,478.97	56,834.24	50,656.61	612,000.00	782,000.00	697,000.00
Geocomposite	13.76	81,544.78	108,726.38	95,135.58	1,122,000.00	1,496,000.00	1,309,000.00
Cover and Vegetative Soil	13.76	32,123.70	64,247.41	48,185.55	442,000.00	884,000.00	663,000.00
Seed, Mulch, and Fertilizer	13.76	2,471.05	4,942.11	3,706.58	34,000.00	68,000.00	51,000.00
Gas Management System	13.76	71,660.57	86,486.89	79,073.73	986,000.00	1,190,000.00	1,088,000.00
Run-Off Control System	13.76	12,355.27	17,297.38	14,826.32	170,000.00	238,000.00	204,000.00
QA/QC	13.76	185,329.06	247,105.41	216,217.23	2,550,000.00	3,400,000.00	2,975,000.00
Total		**560,929.27**	**805,563.63**	**683,246.45**	**7,718,000.00**	**11,084,000.00**	**9,401,000.00**

Table 8.6 Economic Analysis of Landfill: Postclosure and Maintenance Costs (imperial)

Task	Area [acre]	Costs [$/acre] per Year			Total Costs [$] per Year		
		Min.	Max.	Average	Min.	Max.	Average
Security and Fencing Repair	34.00	3.00	6.00	4.50	102.00	204.00	153.00
Cap and Cover Maintenance	34.00	300.00	567.00	433.50	10,200.00	19,278.00	14,739.00
Leachate Machinery Maintenance	34.00	900.00	1,200.00	1,050.00	30,600.00	40,800.00	35,700.00
Landfill Gas Machinery Maintenance	34.00	450.00	570.00	510.00	15,300.00	19,380.00	17,340.00
Wells/Probes	34.00	20.00	30.00	25.00	680.00	1,020.00	850.00
Environmental Monitoring	34.00	450.00	575.00	512.50	15,300.00	19,550.00	17,425.00
Total (per Year)		2,123.00	2,948.00	2,535.50	72,182.00	100,232.00	86,207.00
Total (30 Years)		**63,690.00**	**88,440.00**	**76,065.00**	**2,165,460.00**	**3,006,960.00**	**2,586,210.00**

Table 8.7 Economic Analysis of Landfill: Postclosure and Maintenance Costs (metric)

Task	Area [sq m]	Costs [$/sq m] per Year			Total Costs [$] per Year		
		Min.	Max.	Average	Min.	Max.	Average
Security and Fencing Repair	13.76	7.41	14.83	11.12	102.00	204.00	153.00
Cap and Cover Maintenance	13.76	741.32	1,401.09	1,071.20	10,200.00	19,278.00	14,739.00
Leachate Machinery Maintenance	13.76	2,223.95	2,965.26	2,594.61	30,600.00	40,800.00	35,700.00
Landfill Gas Machinery Maintenance	13.76	1,111.97	1,408.50	1,260.24	15,300.00	19,380.00	17,340.00
Wells/Probes	13.76	49.42	74.13	61.78	680.00	1,020.00	850.00
Environmental Monitoring	13.76	1,111.97	1,420.86	1,266.42	15,300.00	19,550.00	17,425.00
Total (per Year)		5,246.05	7,284.67	6,265.36	72,182.00	100,232.00	86,207.00
Total (30 Years)		**157,381.43**	**218,540.02**	**187,960.73**	**2,165,460.00**	**3,006,960.00**	**2,586,210.00**

8.2 Economic Analysis of WTE

Table 8.8 Economic Analysis of Waste-to-Energy Plant: Assumptions

General	
Price Level Ratio of Purchasing Power Parity Conversion Factor (Croatia)	0.6
Lifetime [years]	35
Yearly Waste Capacity [t]	100,000
Total Waste Capacity (35 Years) [t]	3,500,000
Steam Generator Efficiency [%]	80
Percentage of Plastic in MSW [%]	12.8
Percentage Heat of Produced Energy [%]	75
Percentage Electricity of Produced Energy [%]	25
Energy per Cubic Meter in Gas [kWh]	12.5
Average Lower Heating Value MSW [MJ/kg]	14.98
Average Lower Heating Value Plastic [MJ/kg]	36.16
Cost/Revenue per kWh Electricity [$/kWh]	0.1027
Cost/Revenue per Cubic Feet Gas [$/cubic ft]	0.0039
Cost/Revenue per Cubic Meter Gas [$/cubic meter]	0.1377

Table 8.9 Economic Analysis of Waste-to-Energy Plant: Investment Costs

Type of Costs	Costs Croatia [€]	PP Conversion Factor	Euro-Dollar-Rate	Costs USA [$]
Infrastructure and Waste Storage	4,600,000.00	0.60	1.21	9,296,569.33
Combustion System and Steam Generator	19,500,000.00			39,409,370.00
Water and Steam System	8,000,000.00			16,167,946.67
Design	2,000,000.00			4,041,986.67
Construction	7,000,000.00			14,146,953.33
Electro-Mechanical Installations	5,000,000.00			10,104,966.67
Semi-Dry Treatment	1,200,000.00			2,425,192.00
Bag Filter	2,200,000.00			4,446,185.33
Selective Non-Catalytic Reduction (SNCR) System	800,000.00			1,616,794.67
Other Investment Costs	6,000,000.00			12,125,960.00
Total	**4,200,000.00**			**113,781,924.67**

Table 8.10 Yearly Operating and Maintenance Costs [$]

Personnel Salaries	Number per Shift	Cost per Person per Year [$]	Shifts per Day	Total Costs [$]
Worker	15	30,000.00	3	1,350,000.00
Environmental Engineer	5	90,000.00	3	1,350,000.00
Maintenance Engineer	3	65,000.00	3	585,000.00
Manager (Energy Engineering)	1	95,000.00	3	285,000.00
Machine and Emission Costs	**Costs Croatia [€]**	**PP Conversion Factor**	**Euro-Dollar-Rate**	**Costs USA [$]**
System Maintenance (3% of Invest.)	1,689,000.00	0.60	1.21	3,413,457.74
Process Water	12,000.00			24,251.92
Natural Gas	85,000.00			171,784.43
Reagent for SNCR	80,000.00			161,679.47
Reagent for Semi-Dry Treatment	70,000.00			141,469.53
Bottom Ash-Disposal	1,380,000.00			2,788,970.80
Flying Ash from Steam Generator	138,000.00			278,897.08
Bag Filter Residues	1,575,000.00			3,183,064.50
Emission Fees	13,560.00			27,404.67
Type of Cost	**Consumption per Ton of Waste [MWh/t]**	**Consumption per Year [MWh]**	**Costs per MWh [$/MWh]**	**Total Costs [$]**
Electricity	0.10	10,000.00	102.70	102,700.00
Heat	0.05	5,000.00	11.02	2,754.55
Yearly Operating and Maintenance Costs [$]				**13,866,434.69**
Total Operating and Maintenance Costs (35 Years) [$]				**485,325,214.20**

Table 8.11 Economic Analysis of Waste-to-Energy Plant: Average Lower Heating Value (LHV) of Plastic

Type of Plastic	LHV [MJ/kg]	% in Waste [%]	Total [MJ/kg]
PET	23.2	14.16	3.29
HDPE	44.6	17.39	7.76
PP	42.7	22.62	9.66
LDPE	42.2	22.84	9.64
PVT	19.2	6.64	1.27
PS	42.0	2.71	1.14
Others	25.0	13.63	3.41
Total [MJ/kg]			**36.16**

Table 8.12 Economic Analysis of Waste-to-Energy Plant: Average Lower Heating Value (LHV) of Municipal Solid Waste

Type of Waste	LHV [MJ/kg]	% in Waste [%]	Total [MJ/kg]
Paper	19.12	25.00	4.78
Glass	0.00	4.20	0.00
Metals	0.00	9.40	0.00
Plastics	36.16	13.20	4.77
Rubber and Leather	31.28	3.40	1.06
Textiles	16.05	6.30	1.01
Wood	11.63	6.70	0.78
Food	6.05	15.20	0.92
Yard Trimmings	6.98	13.10	0.91
Other	21.05	3.50	0.74
Total [MJ/kg]			**14.98**

Table 8.13 Economic Analysis of Waste-to-Energy (WTE) Plant: Tipping Fee

State	Number of WTE Plants	Average WTE Tipping Fee [$/t]	Total
Alabama	1	25.00	25.00
Connecticut	7	64.00	448.00
Florida	12	52.92	635.04
Iowa	1	64.00	64.00
Massachusetts	7	69.00	483.00
Minnesota	9	55.00	495.00
New Hampshire	2	69.00	138.00
New Jersey	5	85.00	425.00
New York	10	72.34	723.40
Washington	3	98.00	294.00
Wisconsin	2	51.00	102.00
Total	59		3,832.44
Overall Average Tipping Fee			**64.96**

8.3 Economic Analysis of Recycling

Table 8.14 Economic Analysis of Plastics Recycling: Overall Assumptions

Percentage of PET in Plastic Waste [%]	14.16
Average Price of Recycled PET Pellets [$/lb]	0.58
Price of Recycled PET Pellets [$/kg]	1.26
Electricity Price [$/kWh]	0.1027
Diesel Price [$/gallon]	2.198
Diesel Price [$/l]	0.5807
Water Price [$/gallon]	0.015
Water Price [$/l]	0.0040

Table 8.15 Economic Analysis of Plastics Recycling: Materials Recovery Facility (MRF) Assumptions

Lifetime [years]	10
Processing Building Size (sq. ft)	80,000
Waste Handled per Hour [t]	30
Yearly Waste Handling [t]	120,000
Total Waste Handling (10 years) [t]	1,200,000
Residues Rate [%]	2
Correctness of Manual Plastic Sorting [%]	91
Shifts per Day	2.0
Hours per Shift [h]	8.5
Breaktime per Shift [h]	0.5
Effective Working Time [h]	8.0
Work Days per Week [days]	5.0
Weeks per Year [weeks]	52.0
Yearly Working Hours [h]	4,160.0

Table 8.16 Economic Analysis of Plastics Recycling: Materials Recovery Facility (MRF) Investment Costs

Building and Site	**Initial Investment**
Land [$]	675,000.00
Site Work [$]	720,000.00
Scale House [$]	600,000.00
MRF Building [$]	8,600,000.00
Construction, Planning & Surveying [$]	3,500,000.00
Machines	**Initial Investment**
Metering Bin [$]	150,000.00
OCC Screen [$]	175,000.00
Debris Roll Screen [$]	220,000.00
Newspaper Screen [$]	400,000.00
Polishing Screen [$]	280,000.00
Recycling Magnets [$]	35,000.00
Eddy Current Separator [$]	128,000.00
Optical Plastic Sorting Machine [$]	225,000.00
Baler [$]	550,000.00
Additional Equipment	**Initial Investment**
Conveyor [$]	50,000.00
Rolling Stock [$]	350,000.00
Collection Cars [$]	1,000,000.00
Total Investment Costs [$]	**17,658,000.00**

Table 8.17 Economic Analysis of Plastics Recycling: Material Recovery Facility (MRF) Operating and Maintenance Costs

Personnel Salaries per Year	Number per Shift	Cost per Person per Year [$]	Shifts per Day	Total Costs per Year [$]
Plant Manager	1	89,000.00	2	178,000.00
Operations Foreman	2	55,000.00		220,000.00
Sorters	25	30,000.00		1,500,000.00
Scale House Attendant	2	37,500.00		150,000.00
Equipment Operators	4	35,000.00		280,000.00
Spotters on Tip Floor	2	31,000.00		124,000.00
Marketing Manager	2	50,000.00	1	100,000.00

Facility Costs		Costs per Year [$]
Consumables/Services		28,000.00
Baling Wire		250,000.00
Insurance		45,000.00
Administration		20,000.00

Machines - Operating Costs	Efficiency [%]	Electricity Consumption [kW]	Running Hours per Year [h]	Electricity Costs per kWh [$/kWh]	Yearly Electricity Costs[$]
Metering Bin	100.00	15.00	4160	0.0996	6,215.04
OCC Screen	70.00	8.50			3,521.86
Debris Roll Screen	97.00	30.00			12,430.08
Newspaper Screen	85.00	5.50			2,278.85
Polishing Screen	91.00	10.00			4,143.36
Recycling Magnets	98.00	4.00			1,657.34
Eddy Current Separator	97.00	9.00			3,729.02
Optical PET Sorting Machine	98.00	13.00			5,386.37
Baler	100.00	63.00			26,103.17

Machines - Maintenance Costs		Maintenance Costs per Year [$]
Metering Bin		100.00
OCC Screen		10,000.00
Debris Roll Screen		10,000.00
Newspaper Screen		13,000.00
Polishing Screen		10,000.00
Recycling Magnets		5,000.00
Eddy Current Separator		5,000.00
Optical PET Sorting Machine		5,000.00
Baler		5,000.00

Conveyor - Operating Costs	Efficiency [%]	Electricity Consumption [kW]	Running Hours per Year [h]	Electricity Costs per kWh [$/kWh]	Yearly Electricity Costs[$]
Conveyor	100.00	5.60	4,160	0.1027	2392.50

Conveyor - Maintenance Costs		Maintenance Cost per Year [$]
Conveyor		10,000.00

Rolling Stock - Operating Costs	Efficiency [%]	Diesel Consumption [L/t]	Waste per Year [t]	Diesel Price [$/L]	Yearly Diesel Costs [$]
Rolling Stock	100.00	10.00	120,000	0.7434	892,056.61

Rolling Stock - Maintenance Costs		Maintenance Cost per Year [$]
Rolling Stock		5,000.00

Residues Costs	Percentage of Residues [%]	Residues per Year [t]	Tipping Fee per Ton [$]	Total Costs per Year [$]
Residues Costs per Year	2	2,400	40	96,000.00

Transportation and Collection Costs	Tons Collected per Year	Transportation Costs per Ton [$/t]	Transportation Costs per Year [$]
Transportation and Collection	120,000	36.84	4,420,800.00

Yearly Operating and Maintenance Costs [$]	8,254,465.90
Total Operating and Maintenance Costs (10 Years) [$]	82,544,659.03

Table 8.18 Economic Analysis of Plastics Recycling: Plastic Recycling Facility (PRF) Assumptions

Lifetime [years]	10
Processing Building Size (sq. ft)	70,000
Yearly PET Capacity [t]	15,000
Total Waste Capacity (10 Years) [t]	150,000
Separation Efficiency [%]	91
Dollar - Pound Rate [$/£]	1.46
Shifts per Day	3.0
Hours per Shift [h]	8.5
Breaktime per Shift [h]	0.5
Effective Working Time [h]	8.0
Work Days per Week [Days]	5.0
Weeks per Year [Weeks]	52.0
Yearly Working Hours [h]	6,240.0

Table 8.19 Economic Analysis of Plastics Recycling: Plastic Recycling Facility (PRF) Investment Costs

Building and Site	**Initial Investment [£]**	**Dollar - Pound Rate [$/£]**	**Initial Investment**
Design and Project Management	800,000.00	1.46	1,168,000.00
Civil Engineering	1,200,000.00		1,752,000.00
Land [$]			590,000.00
Site Work [$]			630,000.00
Machines	**Initial Investment [£]**	**Dollar - Pound Rate [$/£]**	**Initial Investment [$]**
Bale Breaker	5,620,000.00		8,205,200.00
Flake Washing Plant (Includes Granulating, Drying, Washing)	9,444,000.00	1.46	13,788,240.00
PET Extrusion	2,643,000.00		3,858,780.00
Metering Bin			150,000.00
Additional Equipment			**Initial Investment [$]**
Rolling Stock [$]			150,000.00
Total Investment Costs [$]			**30,142,220.00**

Table 8.20 Economic Analysis of Plastics Recycling: Plastic Recycling Facility (PRF) Operating and Maintenance Costs

Personnel Salaries per Year	Number per Shift	Cost per Person per Year [$]	Shifts per Day	Total Costs [$]
Sorting and Bale Breaking	6	30,000.00	3	540,000.00
Flake Washing	2	30,000.00		180,000.00
PET Extrusion	2	30,000.00		180,000.00
Waste Processing	2	30,000.00		180,000.00
Equipment Operators	4	35,000.00		420,000.00
Maintenance Staff	2	62,715.00		376,290.00
Plant Manager	1	89,000.00		267,000.00
Marketing Manager	1	50,000.00	1	50,000.00

Machines - Maintenance Costs	Investment Costs [$]	Maintenance [%]	Maintenance Costs [$]
Maintenance Costs of Machines	25,852,220.00	5	1,292,611.00

Machines - Operating Costs	Electricity Consumption [kWh/t]	Electricity Price [$/kWh]	Water Consumption [L/t]	Water Price [$/l]	Tons per Year [t]	Total Costs [$]
Bale Breaker	8.00	0.1027	0.00	0.0040	15,000	12,324.00
Flake Washing Plant	350.00		1.80			539,281.99
PET Extrusion	325.00		0.40			500,686.28
Metering Bin	1.50		0.00			2,310.75

Rolling Stock - Operating Costs	Diesel Consumption [L/t]	Waste per Year [t]	Diesel Price [$/L]	Yearly Diesel Costs [$]
Rolling Stock	10.00	15,000	0.7434	111,507.07

Yearly Operating and Maintenance Costs [$]	**4,652,011.09**
Total Operating and Maintenance Costs (10 Years) [$]	**46,520,110.91**

8.4 Optimization I: Reduction of Sorting Processes

Table 8.21 Economic Analysis of Optimization I: Assumptions

Percentage of PET in Plastic Waste [%]	14.16
Average Price of Recycled PET Pellets [$/lb]	0.58
Price of Recycled PET Pellets [$/kg]	1.28
Electricity Price [$/kWh]	0.1027
Diesel Price [$/gallon]	2.814
Diesel Price [$/l]	0.7434
Water Price [$/gallon]	0.015
Water Price [$/l]	0.0040
Lifetime [years]	10
Processing Building Size (sq. ft)	80,000
Yearly Working Hours [h]	6,240
Yearly Plastic Waste Handling [t]	100,000
Total Plastic Waste Capacity (10 Years) [t]	1,000,000
Yearly PET Handling [t]	15,000
Total Waste Capacity (10 Years) [t]	150,000
Separation Efficiency [%]	91
Dollar - Pound Rate [$/£]	1.46
Shifts per Day	3.0
Hours per Shift [h]	8.5
Breaktime per Shift [h]	0.5
Effective Working Time [h]	8.0
Work Days per Week [days]	5.0
Weeks per Year [weeks]	52.0
Yearly Working Hours [h]	6,240.0

Table 8.22 Economic Analysis of Optimization I: Additional Investment Costs

Investment Costs Additional Building and Site	**Initial Investment [$]**
Additional Land [$]	85,000.00
Additional Site Work [$]	90,000.00
Scale House [$]	600,000.00
Additional Building [$]	1,000,000.00
Investment Costs Additional Machines	**Initial Investment [$]**
Metering Bin [$]	150,000.00
Optical PET Sorting Machine [$]	225,000.00
Baler [$]	550,000.00
Investment Costs Additional Equipment	**Initial Investment [$]**
Conveyor [$]	50,000.00
Additional Rolling Stock [$]	200,000.00
Collection Cars [$]	1,000,000.00
Total Additional Investment Costs [$]	**3,950,000.00**

Table 8.23 Economic Analysis of Optimization I: Additional Operating and Maintenance Costs

Personnel Salaries per Year	**Number per Shift**	**Cost per Person per Year [$]**	**Shifts per Day**		**Total Costs [$]**
Sorters	6	30,000.00	3		540,000.00
Scale House Attendant	2	37,500.00			225,000.00
Equipment Operators	1	35,000.00			105,000.00
Spotters on Tip Floor	1	31,000.00			93,000.00
Facility Costs per Year					**Costs per Year [$]**
Baling Wire					250,000.00
Machines - Operating Costs	**Efficiency [%]**	**Electricity Consumption [kW]**	**Running Hours per Year [h]**	**Electricity Costs per kWh [$/kWh]**	**Yearly Electricity Costs[$]**
Metering Bin	100.00	15.00	6,240	0.1027	9,612.72
Optical PET Sorting Machine	98.00	13.00			8,331.02
Baler	100.00	63.00			40,373.42
Machines - Maintenance Costs					**Maintenance Costs per Year [$]**
Metering Bin					100.00
Optical PET Sorting Machine					5,000.00
Baler					5,000.00
Rolling Stock - Operating Costs	**Efficiency [%]**	**Diesel Consumption [L/t]**	**Waste per Year [t]**	**Diesel Price [$/L]**	**Yearly Diesel Costs [$]**
Rolling Stock	100.00	10.00	100,000	0.7434	743,380.51
Rolling Stock - Maintenance Costs					**Maintenance Costs per Year [$]**
Rolling Stock					5,000.00
Transportation and Collection Costs	**Tons per Year [t]**		**Transportation Costs per Ton [$/t]**		**Transportation Costs per Year**
Transportation and Collection	100,000		36.84		**3,684,000.00**
Yearly Operating and Maintenance Costs [$]					5,713,797.67
Total Operating and Maintenance Costs (10 Years) [$]					**57,137,976.74**

8.5 Optimization II: Upcycling of Plastic Waste by Blending

Table 8.24 Economic Analysis of Optimization II for Materials Recovery Facility (MRF): Additional Costs

Additional Investment Costs					**Initial Investment [$]**
LDPE-PP Optical Sorter					900,000.00
Additional Operating Costs	**Efficiency [%]**	**Electricity Consumption [kW]**	**Running Hours per Year [h]**	**Electricity Costs per kWh [$/kWh]**	**Yearly Electricity Costs [$]**
LDPE-PP Optical Sorter	95.00	40.00	4160	0.1027	17,089.28
Additional Maintenance Costs					**Maintenance Costs per Year [$]**
LDPE-PP Optical Sorter					10,000.00
Overall Additional MRF Costs [$]					**1,170,892.80**
Additional MRF Costs per Ton [$/t]					**0.98**

Table 8.25 Economic Analysis of Optimization II for Plastic Recycling Facility (PRF): Additional Investment Costs

Building and Site	**Initial Investment [£]**	**Dollar - Pound Rate [$/£]**	**Initial Investment [$]**
Design and Project Management	500,000.00	1.46	730,000.00
Land [$]	590,000.00		
Site Work [$]	630,000.00		
Machines	**Initial Investment [£]**	**Dollar - Pound Rate [$/£]**	**Initial Investment [$]**
Bale Breaker	5,620,000.00	1.46	8,205,200.00
Flake Washing Plant (Includes Granulating, Drying, Washing)	9,444,000.00		13,788,240.00
LDPE-PP Extruder	1,422,000.00		2,076,120.00
Metering Bin	150,000.00		
Additional Equipment			**Initial Investment [$]**
Rolling Stock			50,000.00
Total Investment Costs [$]			**26,069,560.00**

Table 8.26 Economic Analysis of Optimization II for Plastic Recycling Facility (PRF): Additional Operating and Maintenance Costs

Personnel Salaries per Year	Number per Shift	Cost per Person per Year [$]	Shifts per Day	Total Costs [$]
Sorting and Bale Breaking	6	30,000.00		540,000.00
Flake Washing	2	30,000.00		180,000.00
LDPE-PP Extrusion	2	30,000.00	3	180,000.00
Waste Processing	2	30,000.00		180,000.00
Equipment Operators	4	35,000.00		420,000.00

Machines - Maintenance Costs	Investment costs [$]	Maintenance [%]	Maintenance Costs [$]
	24,069,560.00	5	1,203,478.00

Machines - Operating Costs	Electricity Consumption [kWh/t]	Electricity Price [$/kWh]	Water Consumption [L/t]	Water Price [$/l]	Tons per Year [t]	Total Costs [$]
Bale Breaker	8.00		0.00			12,324.00
Flake Washing Plant	350.00	0.1027	1.80	0.0040	15,000	539,281.99
LDPE-PP Extrusion	275.00		0.40			423,661.28
Metering Bin	1.50		0.00			2,310.75

Rolling Stock - Operating Costs	Diesel Consumption [L/t]	Waste per Year [t]	Diesel Price [$/L]	Yearly Diesel Costs [$]
Rolling Stock	10.00	15,000	0.7434	111,507.08

Yearly Operating and Maintenance Costs [$]	**3,792,563.09**
Total Operating and Maintenance Costs (10 Y.) [$]	**37,925,630.91**

Index